GROUP SELECTION

GROUP SELECTION

Edited by

GEORGE C. WILLIAMS

AldineTransaction
A Division of Transaction Publishers
New Brunswick (U.S.A.) and London (U.K.)

New paperback printing 2008

This book is printed on acid-free paper that meets the American National Standard for Permanence of Paper for Printed Library Materials.

Library of Congress Catalog Number: 2008016639
ISBN: 978-0-202-36222-9
Printed in the United States of America

Library of Congress Cataloging-in-Publication Data

Group selection / George C. Williams, editor.
p. cm.
Includes bibliographical references and index.
ISBN 978-0-202-36222-9
1. Group selection (Evolution) 2. Animal populations. I. Williams, George C. (George Christopher), 1926

QH366.2.W55 2008
576.8'2—dc22 2008016639

ACKNOWLEDGMENTS

The editor wishes to thank the authors of the works included in this volume for permission to use their writings for a purpose somewhat different from their original intent. I am also grateful to the following publishers for permission to reprint their copyrighted materials: Academic Press, publisher of the *Journal of Theoretical Biology;* Cambridge University Press, publisher of the *Annals of Human Genetics;* the University of Chicago Press, publisher of the *American Naturalist;* and Macmillan (Journals) Limited, publisher of *Nature.*

Contents

GROUP SELECTION

Introduction

GEORGE C. WILLIAMS

A robin's serenade when day is through

And there's a bluebird
Singin' to his lady love
Above, a love song. . . .

OLD SONGS

Demon or bird! (said the boy's soul,)
Is it indeed toward your mate you sing?

WALT WHITMAN

As we all know, living things are engaged in a struggle for existence. From molecular interactions to the geometry of a turtle shell we see variations on the theme of self-preservation. The more we study living nature, the more we marvel at the ingenious devices by which animals and plants win the necessities of life and avoid destruction. When an organism shows some property that is not immediately intelligible in relation to continued survival, later developments will often provide the missing parts of the story. Physiologists did not at first understand the blood. When advances in chemistry and microscopy permitted, its role in cellular respiration and nutrition and its precise adaptation to this role were made clear. The blood was understood as a vital cog in the machinery of self-preservation. Physiologists continue to find that postulating answers to the question: "How might this contribute to survival?" can provide fruitful insights.

Fruitful, but not infallible. Any serious observer of nature will soon turn up phenomena that have no relation to survival, or that seem to be clear violations of the principle of self-preservation. This often tends to be true of ways in which organisms interact with one another. Bird songs are a good example. A bird sings, and this singing may influence the behavior of other birds of the same species, but neither the singing nor the reactions to it seem to contribute in any way to survival. If anything, we might envision the opposite effect. The woods are full of things that think birds are delicious. Why should a bird make itself so conspicuous and easy to find? The obvious answer is found in the excerpts from the verses quoted above. The bird sings to win a mate, without which he cannot reproduce. Reproduction provides the clearest and most conspicuous exceptions to the otherwise pervasive principle of self-preservation.

Darwin's great achievement was to show that both the mechanisms of self-preservation and those of reproduction are explained by a more basic principle of natural selection, the reproductive survival of the fittest. All through their history the robins that were best able to avoid destruction and grow into vigorous adults that were then best able to produce and rear offspring were the ones that had their characteristics best represented in the next generation. Evolution is expected in this way to produce organisms that are optimally designed for their own reproduction. It favors self-preservation only because an organism that does not survive will not reproduce.

This theory of natural selection gives an objective view of the nature of biological organization, and a majority of modern biologists, myself included, accept this view. This does not mean that we think Darwinian selection provides immediate and complete answers to all questions. The theory enables us to explain why the stickleback in a pond will obey the law of self-preservation in most respects, but will take great personal risks in courtship and the defense of its nest full of young. We are less immediately successful with the problem of why the stickleback works so hard at reproduction, while the killifish in the same pond merely abandons the eggs as soon as they are fertilized. In such cases the assumption is made that careful study will provide the answer. We would expect to find that their different structures and circumstances and approaches to the problems of life make the killifish way of reproducing best for the killifish, and the stickleback way best for the stickleback.

Altruistic Behavior

We will return later to the courtship and related activities of the robin and other song birds, but I want first to consider another kind of vocalization. In the winter, when they are not breeding, the robins move about in small flocks in the warmer parts of the country. When one of the members of the flock is frightened by a predator, it may utter a special chirp, which instantly alerts other members of the flock to the danger. This special chirping behavior had to be acquired by the evolutionary process. For this to happen it was necessary, on the usual view of natural selection, that individuals with a greater than average tendency to chirp should survive and reproduce more than those in which this tendency was less developed. The chirp shows no special relation to reproduction. It is heard both in the breeding season and out. Must it therefore contribute to survival, and if so, how? There is no problem in the reaction to the chirp. Those that recognize it as a danger signal and behave appropriately may well avoid sudden death. But what good does it do the bird that gives the signal? As with prenuptial singing, we might imagine that by sounding the alarm the bird would call attention to itself and thereby increase its own danger. However, suppose it failed to signal, and merely took flight and saved itself without concern for others. Occasionally, at least, this would result in one of the others being caught and killed, when it might have escaped if it had been warned. So what? If this had any effect on the first bird, it would be a beneficial one. It would mean one less competitor for food and other necessities. Yet somehow evolution has run counter to expectation here. The warning note was evolved and continues to persist.

Many birds and other animals have alarm calls or other signals by which they warn one another of danger. In every case it would seem that one individual expends effort and perhaps increases its own risk to provide a benefit to some other individual in its social group. It was this kind of altruistic behavior that first suggested the idea of group selection. The natural selection of individuals can produce only mechanisms of individual advantage. To produce mechanisms of group benefit, that depend on individual self-sacrifice, a natural selection of groups must be operating.

The most extreme example of altruistic behavior, and therefore crucial evidence on group selection, is seen in the social insects, in which one individual, a worker or soldier, entirely refrains from breeding and devotes its life to assisting the queen or royal couple and their young. How could evolution produce this sort of thing if it is always guided by the reproductive survival of the fittest? Is it not a contradiction to say that selection for effective reproduction can result in increasing numbers of sterile individuals? Do not bird flocks and especially the social insects provide incontrovertible evidence for group selection?

For a while this seemed the obvious conclusion, but I think it is fair to say that most of the serious recent thinking on this subject indicates that altruistic behavior as seen in bird flocks and even in the social insects may be attributed to the traditional kind of natural selection acting on individuals. A key to the problem is seen in the logical inference that altruism, to the extent that it normally benefits close relatives, will be favored in evolution. There is no problem in understanding altruism directed at offspring, because it is through them that one ordinarily achieves representation in the next generation. As the first of the papers by Hamilton points out, offspring are merely one kind of relative. Inasmuch as one's brother or sister has much the same genetic makeup as oneself, his survival and reproduction is partially equivalent to one's own survival and reproduction. The same is true, to a rapidly decreasing extent, of more distant relatives. It is formally reasonable (but not necessarily convincing to everyone) that the giving of a warning call by a robin in the winter would be favored by selection if the flocks are often made up of close relatives. In this way a theory of "kin selection" can explain at least some of the examples of altruism in nature.

Another possibility, which I believe deserves more attention than it has received, is that some altruistic traits arise by accident or default. The alarm calls of birds do work in the breeding season too, and they may have been evolved for the protection of offspring. That they persist between breeding seasons may merely mean that no mechanism has evolved to turn them off when not needed. This would imply that the disadvantage to the caller is extremely small, which might be true for the robin's alarm note, although it is certainly not true of the sterility of a worker bee.

Unfortunately, it is unlikely that really decisive evidence for choosing between group selection and kin selection, either in general or even for specific examples, will be forthcoming in the near future. It is difficult to determine the degree to which nonbreeding social groups are made up of close relatives, and almost impossible to measure the cost of altruism to the donor or the benefit to the recipient.

The Regulation of Population Density

In recent years the focus of the group selection controversy has shifted away from consideration of overtly benevolent interactions among individuals. In the current decade the emphasis is on a more subtle form of altruism, in which long-term survival of the group may depend more on a certain degree of antagonism and interference than on mutual aid. Specifically the interference would be with the reproductive process, and it would act as a socially regulated birth control program. Something of the sort has been hypothesized by a number of workers, but none so effectively as Wynne-Edwards.

The basic question posed by Wynne-Edwards is this: Is the reproductive physiology and behavior seen in animal populations best regarded as a way of ensuring the survival of the species, or as the attempt by each individual to maximize its own reproductive survival, regardless of consequences to the population as a whole? In answering this question, Wynne-Edwards looks first for what could be regarded as the most important hazard to the long-term survival of a wild population. He sees overexploitation of resources as this greatest of problems to be overcome.

It is a matter of simple arithmetic that the reproductive capabilities of even the slowest breeding species are capable of producing astronomical numbers in a surprisingly short time. It is a matter of simple observation that this does not occur, and it is obvious why it cannot. Animals need food and other necessities to survive, and these are always in limited supply. We might therefore conclude that every population will be as large as the environmental resources permit, and that shortages of food or other resources are what prevent a population of animals from increasing to greater numbers.

Yet one may get the impression that this is not what is really going on. The woods may be teeming with the insects on which a robin feeds, and it seems unlikely that a few more robins would make an appreciable difference, yet the robins fail to increase in number. An insect on which the birds feed may eat the leaves of a certain tree. These leaves may be there in enormous numbers, seemingly enough to support a much larger population of the insect, yet the insect fails to increase and make use of the apparent abundance.

Wynne-Edwards proposes that animal populations are in fact normally smaller than they might be, smaller than resources would temporarily permit. He ascribes this situation to population control mechanisms operative in the animals themselves. He says, in effect, that animals as a very general rule practice birth control to an effective degree.

Wynne-Edwards sees an important analogy in man's exploitation of renewable resources. It is often true that a fishery or other exploitative industry, in which each participant tries always to maximize his immediate profit, will sooner or later collapse. The reason is that there is no easy way of telling when the industry is merely consuming some of the productivity of the exploited population and when it is consuming part of the capital on which later productivity must depend. Even if each individual or company knew when the danger point was reached, there would be no point in limiting one's own catch unless everyone else did the same. Overfishing continues until the prey species is reduced to the point where it is no longer profitable to pursue, and there is wholesale bankruptcy of the industry. Wynne-Edwards believes that analogous disasters would arise in nature if populations did not actively keep their consumption, and therefore their numbers, at a safe level. In his view, the primary reason that animals do not eat themselves to near or complete extinction is that they impose restraints on their own reproduction.

Territoriality

The theory of adaptive regulation of population size, like other important ideas in science, could have been formulated only after certain earlier developments had prepared the way. In this case the most important prior discovery may have been territoriality. At this point we might take a closer look at the bird songs. The obvious interpretation, implied by calling the robin's song a serenade, turns out to be less obvious and more complicated when we give it some critical attention.

First, the males may reach the nesting areas a week or so before the females and start singing right away, long before any females could possibly hear them. The courtship idea might conceivably be rescued if the male needs to practice first. Don Giovanni must have practiced his serenade in private before using it on Donna Elvira's maid. Unfortunately the analogy is false. A robin sings beautifully the first time he tries, and veterans of several seasons sing as early as those newly matured. Careful observation forces the conclusion that the early songs are directed not at potential mates but at rival males. Each robin has a proprietary interest, not only in the nest and its immediate vicinity, but in a large surrounding area. The singing functions to discourage other males from establishing themselves within an area that the singer regards as his own. This area is called a *territory* by students of animal behavior, and a territory, by definition, is an area that one individual defends against intrusion by other members of its species.

In small birds generally, contested territories and boundary disputes are settled by song contests, and actual fighting, which may occasionally occur, is normally not resorted to. Territories are usually well established and controversies resolved before the females make their appearance. Only then can song take on an additional function in courtship. Perhaps it is something of a disillusionment to learn that the harbingers of spring are not singing love songs; nor are they singing for the poet, as Whitman suggested; they are shouting threats at each other.

It is easy to see the necessity of winning a mate and to understand why courtship may be produced by natural selection. It is less immediately obvious why a territory should be necessary. Reasons may and have been suggested. It may be advantageous to keep other individuals from feeding in one's own territory. This may assure a greater supply of food for oneself and one's offspring. Keeping rival males away may prevent interference with courtship. For the same reasons it may be advantageous to avoid settling in an area that is already crowded with defended territories. These and other explanations that seek the function of territoriality at the individual level have evidence in their favor, evidence that convinces me that territoriality can be explained by selection at the individual level, but not everyone is convinced, because there are some troublesome details. The territories often seem a bit larger than should be necessary for a hunting preserve or for sexual privacy.

Also, they are vigorously defended only against other members of the same species, while other competitors for food are largely ignored. Answers to these problems can be offered, but many biologists remain skeptical. They feel that territoriality and other social phenomena often show features that cannot be reconciled with the view that every individual is maximally adapted to secure his own reproduction.

Wynne-Edwards is one of the skeptics, and he has proposed, and documented in great detail, the alternative view of territoriality as a population control mechanism. He regards territoriality as a way of dividing the resources of the breeding habitat into a limited number of holdings, each with at least a minimum supply of vital resources. Each bird instinctively accepts the constraint: "if I do not have a territory with at least a certain size and quality, I will not attempt to reproduce." Thus there will not ordinarily be more pairs of reproducing robins in an area than there are territories of a certain minimal size. With a ceiling placed on the number of reproducing pairs, it seems reasonable to assume that a brake is placed on the growth of the population.

The minimum territory size will vary with the richness of the habitat and cannot be expected to be quantitatively exact. The breeding of twelve pairs of robins on a farm this year and only ten next year would not be evidence against Wynne-Edwards' theory. (Perhaps great fluctuations, with no indication of a ceiling, would be contrary evidence.) It is really only necessary that the smaller the territory the more vigorously a bird should resist further reduction in its size, so that a point would be reached at which it would be very difficult for a new male to establish a territory without displacing an old one.

A self-imposed limit to increase is most easily envisioned for a species that is monogamous and territorial. This is true of many birds and mammals and lower vertebrates, and of some invertebrates. But Wynne-Edwards does not restrict his theory to these species. He argues for other socially imposed reproductive restraints in species with different kinds of population structures, for instance, gregarious and polygynous species.

All he requires is that there be some way for an animal to find out how crowded it is, and to react to a certain level of crowding by refraining from reproduction, even when successful reproduction would be quite within its physiological capabilities.

Group Selection

Two kinds of criticism have been directed at this theory of birth control in the animal world. One is based on reinterpretation of evidence and says that wild animals do not behave as the theory says they do, or that their behavior would not regulate populations as the theory maintains. It is in the gathering and interpretation of evidence of this sort that the issue must ultimately be decided. The second criticism is that evolution cannot produce self-imposed restraints on reproduction, except perhaps when it is favored by kin selection, as in the social insects. This criticism is based on a rejection of group selection as a force that can counter selection at the individual level.

It was in anticipation of this second criticism, and from a conviction that it was fundamentally in error, that Wynne-Edwards gave considerable attention to the operation of group selection. The idea was not new, as is apparent from the paper by Sturtevant, but Wynne-Edwards discussed its operations and implications in much greater detail than anyone before. The theory regards evolution as proceeding not merely by the differential survival and reproduction of individuals, but also by the differential survival and reproduction of local populations of a species. Just as selection among individuals can produce mechanisms of self-preservation, selection among alternative groups should produce mechanisms of group-preservation. Evolution would proceed by the best adapted groups surviving and replacing the less well endowed, and in turn giving rise to other local populations of which the fittest would be most likely to endure, and so on.

All this seems reasonable at first glance, and just what we would expect to be going on in nature. Indeed, no one could reasonably deny that extinction at all levels, from the local population to such major groups as the dinosaurs, must be a powerful factor in evolution. The difficulty lies not in accepting the reality of the process but in believing that it can somehow prevent a species from evolving in a direction favored by selection at the individual level. Only if it could force evolution counter to the direction of individual selection, would it be capable of producing self-imposed limits on reproduction.

Both theory and criticism can be made clear by an example. Suppose that in a certain city park, of a sort suitable for the nesting of robins, the population is being regulated in the manner envisioned by Wynne-Edwards. We will assume that it is a self-contained population in that the same adults nearly always return to this park year after year, and young produced in the park return to it when it is time for them to reproduce. Because it is effectively regulated in relation to the food resources provided by the park, this population will never overexploit its resources and is therefore likely to survive for a long time. If other populations (in other parks) die out, this one will provide a stock from which the species can be re-established in other areas.

Therefore things are going well with the robins in our park, but suppose an individual appears in which the mechanisms of reproductive restraint are absent or less well developed than is normal. In times of increased numbers, when many of the robins are dutifully neglecting to raise a family, the abnormal bird will be among the active breeders. It will produce more than its fair share of offspring, compared to the normal birds, and the behavioral abnormality will be more abundant in the next generation. When that generation reproduces, the same thing will happen again. Selection at the individual level will cause the population to evolve a decreased level of control on its numbers. This will be true not only in one park but in all of them, and even the best controlled populations would be evolving a loss of control. Selection among populations cannot cause evolution to go in one direction, when each of the populations is evolving in the opposite direction.

Not everyone regards this difficulty as fatal to the theory, and some suggestions have been made that may provide a way out. Nevertheless, I think that even the most enthusiastic proponents of group selection would admit that the logical basis of their theory is not as firm as might be desired, and that this is a matter seriously in need of attention. A rigorously reasoned theory of group selection would presumably provide a more exact idea of the sort of evidence we should look for in work with the real world of animal populations.

Sex

As indicated above, I see in recent years a change in discussions of group selection and possible mechanisms of long-term group benefit. The emphasis at first was on seemingly altruistic behavior. Because of Wynne-Edwards' work the emphasis today is on the regulation of population density. Prophesy is a hazardous exercise in science, but I will venture to suggest that in the future the controversy will center increasingly on the phenomenon of sexuality. The focus of attention will be on the origin and maintenance of meiosis as a part of the reproductive process.

Almost universally, opinion has been that the significance of sexuality, which assures a shuffling of hereditary factors in lines of descent, is that it refines and facilitates the process of evolution. Without sexual recombination, evolution would go more slowly and could probably not have produced the observable diversity and complexity of animals and plants. It aids evolution by providing a way for genes in one line of descent to be combined with those in other lines to produce new combinations to be tested by selection. As indicated in the papers by Smith and by Crow and Kimura, opinion varies on the magnitude of the beneficial effects and on the circumstances under which they are important, but there is near unanimity on the point that sexuality functions to facilitate long-range evolutionary adaptation, and that it is irrelevant and even detrimental to the reproductive interests of an individual.

In other words, sexuality, a phenomenon of near universality and paramount importance, is exactly the sort of thing for which group selection must be postulated, although the force of this conclusion is only slowly coming to be realized. The reason why group selection is indicated here, or even demanded, is clear when we consider what happens to individual reproductive interests when, in meiosis, the number of chromosomes and constituent genes is reduced by half. Each resulting gamete, and zygote that is formed by fertilization, will have a sampling of half the genes of the individual that provides the gametes. In the usual mitotic divisions, each resulting cell preserves the entire genome intact.

Suppose there were two kinds of females in a population; one produced monoploid, fertilizable eggs, and the other skipped meiosis and produced diploid eggs, capable of development without fertilization, and each with exactly the mother's genetic makeup. These parthenogenetic eggs would each contain twice as much of the mother's genotype as is present in a reduced and fertilized egg. Other things being equal, the parthenogenetic female would be twice as well represented in the next generation as the normal one. In a few generations, meiosis and sexual recombination should disappear and parthenogenesis become the normal reproductive pattern. Males would no longer occur.

Meiosis is therefore a way in which an individual actively reduces its genetic representation in its own offspring. Any success that these offspring achieve is shared equally by the mother and the father. The parthenogenetic female shares her reproductive success with no one. Sexual reproduction is analogous to a roulette game in which the player throws away half his chips at each spin. The game is fair as long as everyone behaves in this way, but if some do and some don't, the ones who keep their chips have an overwhelming advantage and will almost certainly win.

Sex has been recognized as an adaptation for long-term group benefit for perhaps forty years, and the existence of such adaptations is formally incompatible with the accepted picture of evolution. Yet only after group-related adaptation had been recognized in social behavior, and a controversial theory proposed to explain it, did the paradox of sexual reproduction begin to be recognized as relevant to the problem. This sequence of developments will someday be recognized as a curious feature of the history of biological thinking in the twentieth century.

Works Included in this Volume

Our intention is to give a candid view of a live issue. The topic chosen is of broad importance and it is one in which there is a serious difference of opinion among leading investigators. There are distinguished biologists who after years of study regard group selection as a force of profound importance in the evolution of a species or more inclusive taxon. To other equally distinguished biologists it is a preposterous aberration which can serve only as a distraction and impediment to our understanding of living organisms. The strength of a scientist's convictions on such issues is seldom explicit in his technical publications. He sees his job as the cautious and impartial marshaling of evidence and closely reasoned arguments, and he seldom spends much eloquence in stating broad conclusions. A reader should bear this in mind as he reads this book.

It need hardly be stated that it was difficult to decide what papers to include and what could be left out. The task was simplified by the fact that constructive and deliberate discussions of group selection as a theory are rather few. It was made difficult by the diversity of topics that are relevant to the controversy. Behavioral and ecological field studies, laboratory investigations of social behavior or experimental populations, and purely abstract treatises on ecology and population genetics are all pertinent. We hope that the papers selected will convey a reasonably balanced picture of the current debate and that the supplementary bibliography will provide a good start for any reader wishing to pursue the topic further.

We have not ignored the virtue of readability, but have given it a lower value than that of fidelity to the issues. Consequently much of what follows will require considerable effort from even a well-prepared reader who wants a thorough understanding of the controversy. I am sure that some of the discussion will be understood in detail by only a few people. However, a less than perfect comprehension can be of great value, and a reader not thoroughly versed in biological and mathematical technicalities should not be discouraged from reading merely because a paper contains unfamiliar words or mathematical notation. One can get an intuitive feel for a line of reasoning without following all the mathematics with which it is supported, although, of course, the understanding would be greatly enriched by careful step-by-step analysis. Similarly it would help to know, in reading Hamilton's work, just what kind of creature a *Polistes* wasp is, its appearance, relationships, distribution, and so on, but this is not essential. All that is directly of interest to the discussion are certain aspects of its social behavior and population structure, and these are adequately summarized by Hamilton.

The papers are reproduced exactly as they appeared in the technical literature, except for deletion of literature references, footnotes, and appendixes. The reader should always assume that statements of fact were adequately documented by the authors in the original publications. These originals are, of course, available in research libraries.

I

Altruistic Behavior and Social Organization

Although its reasoning and statement of conclusions may now be of mainly historical content, the paper by Sturtevant, one of the founders of modern genetics, gives a clear statement of the concept of group selection and its relation to the evolution of altruistic behavior.

Hamilton's work, despite the difference in style, may be considered an extension of Sturtevant's, although Hamilton is obviously not enthusiastic about group selection. It will be difficult for many readers to follow the mathematical development that Hamilton uses, but its outcome, the concept of inclusive fitness, is easy to understand in a qualitative way. There should be no difficulty in understanding its relevance to ensuing discussions of social behavior.

Hamilton provides the fullest development to date of the theory of kin selection as an alternative to group selection. The contribution as a whole—in its artful development of the theoretical model and its fair and thoughtful testing of the model against the pertinent phenomena—can serve as a model of what theoretical biology ought to be like.

1: *On the Effects of Selection on Social Insects*

A. H. STURTEVANT

There are two conflicting tendencies in recent developments of the mathematical theory of the behavior of wild populations. On the one hand there is the view that selection is ever present and is the only effective agent in altering the constitution of populations. On this view selection operates only on individuals, and the members of a group are not thought of as occurring in semi-independent subgroups. On the other hand there is the point of view expressed by Wright, according to which the most favorable conditions for evolution are those in which the members of a large population do not interbreed at random, but exist in more or less independent subgroups, which undergo exchanges of genes at times but are also effectively in competition with each other. On this latter view there is an opportunity, lacking on the first view, for selection to bring about the establishment of characteristics favorable to the group but unfavorable to the individual.

From the *Quarterly Review of Biology*, 13 (1938), 74–76.

The social insects, such as termites, ants, and many bees and wasps, furnish examples in which such characters unfavorable to the individual have in fact become established. In terms of natural selection, a "favorable" character is of course to be taken as meaning a character that leads to the production of more descendants. The sterile castes of the insects named have, therefore, developed a character that is unfavorable by definition. It is sometimes argued that in this special case the rule still holds, since here the colony, rather than the individual insect, is the unit in terms of natural selection, and it may be supposed that the colony produces more offspring as a result of the division of labor associated with the sterile castes. It is clear, however, that all the social insects have arisen from solitary forms in which the sterile caste was absent. It follows that evolution must have resulted in an increase of sterile individuals. At some point in the history of the race there must have been a change from the individual to the colony basis of selection. Unless this change be supposed to have been a sharp one and to have been associated from the first with the necessary genetic adjustments, there must have been an intermediate stage in which some element other than the strict operation of reproductive selection was effective.

In fact, such intermediate stages still exist, and are familiar to all students of the social insects. In order that the colony be the unit in terms of natural selection, and that the existence of the sterile caste offer no difficulties to the advocates of pure selection, it is necessary that each colony have a single fertile queen and that the sterile individuals all be closely related to the queen (presumably her offspring). This is the standard account for the honey-bee, though even here it is clear that workers may at times produce offspring; in other social forms wide deviations from this condition are frequent.

In what follows I shall confine myself to the ants, though evidently somewhat similar relations occur in certain of the termites at least. For detailed summaries of the life-histories and for bibliographies the reader is referred to Wheeler and to Donisthorpe. Fertile workers do not appear to be rare among the ants; usually they produce only male offspring, but there are some satisfactory records of female offspring—the latter presumably due to failure of chromosome reduction in parthenogenetic eggs, since it seems clear that such fertile workers do not mate. The production of males is, however, enough to put these workers into reproductive competition with the queens; and therefore to upset the view that the colony is the unit of selection, to the exclusion of individual selection within the colony. Even more to the point is that, in many ants, more than a single fertile queen is present in a given colony. In such dominant genera as *Myrmica*, *Crematogaster*, and *Tapinoma*, and in the *rufa* and *sanguinea* sections of the genus *Formica*, flourishing nests regularly contain many fertile queens. In some (probably in all) these cases the colony is initiated by a single queen, but newly fertilized queens are adopted by established colonies after the mating flight occurs. In the case of *Formica rufa* at least, there is evidence that these adopted queens need not have been produced in the nest that adopts them—they may even belong to a distinct variety. Here and in *F. exsectoides* it seems probable that the colony long outlives the queen that originally established it, its fertility being dependent upon a whole series of queens that have no necessary genetic relationship to each other or to the founder.

Under these conditions there is evidently an opportunity for individual selection. A queen that produces unusually efficient worker offspring will have to share the benefits of their activities —i.e., her fertile offspring will have no advantage over those of the other queens living in the same nest. A queen that produces a relatively high proportion of fertile offspring will leave more descendants—at the expense of the nest economy, and in the long run to the detriment of the species. This possibility seems to constitute a serious danger to such ants, and may therefore be examined more closely.

There are ants in which the worker caste has disappeared entirely (*Anergates*, *Wheeleriella*, etc.), so that mutations of this sort are evidently possible. If such a one occurs, then a queen of the new type that is adopted by a normal colony will thrive at the expense of the worker offspring of her foster-sisters, but will produce more than her share of the sexual forms arising in the colony. Such a gene may be expected to spread through the population rapidly at first, but must soon be checked by the occurrence of too high a proportion of sexual forms to be supported by the reduced number of workers that will be available. It seems clear that the net result must be a decrease in the efficiency of the colonies and therefore of the total population of the species in the affected area. This will, evidently, leave such a region likely to be invaded by members of the species coming from places where such a process has not happened to occur.

Selection must, then, be thought of as operating, in these forms, on at least three different levels: on the individuals, on the colonies, and on the populations within an area. The interrelations between these three are obviously complex. Similar distinctions apply in the case of another social animal—namely man. Here, of course, the situation is even more complex, and I do not have the intimate knowledge of the data necessary for a profitable discussion.

2 : The Genetical Evolution of Social Behavior. I

W. D. HAMILTON

A genetical mathematical model is described which allows for interactions between relatives on one another's fitness. Making use of Wright's Coefficient of Relationship as the measure of the proportion of replica genes in a relative, a quantity is found which incorporates the maximizing property of Darwinian fitness. This quantity is named "inclusive fitness." Species following the model should tend to evolve behavior such that each organism appears to be attempting to maximize its inclusive fitness. This implies a limited restraint on selfish competitive behavior and possibility of limited self-sacrifices.

Special cases of the model are used to show (a) that selection in the social situations newly covered tends to be slower than classical selection, (b) how in populations of rather non-dispersive organisms the model may apply to genes affecting dispersion, and (c) how it may apply approximately to competition between relatives, for example, within sibships. Some artificialities of the model are discussed.

1. Introduction

With very few exceptions, the only parts of the theory of natural selection which have been supported by mathematical models admit no possibility of the evolution of any characters which are on average to the disadvantage of the individuals possessing them. If natural selection followed the classical models exclusively, species would not show any behavior more positively social than the coming together of the sexes and parental care.

Sacrifices involved in parental care are a possibility implicit in any model in which the definition of fitness is based, as it should be, on the number of adult offspring. In certain circumstances an individual may leave more adult offspring by expending

From the *Journal of Theoretical Biology*, 7 (1964), 1–16.

care and materials on its offspring already born than by reserving them for its own survival and further fecundity. A gene causing its possessor to give parental care will then leave more replica genes in the next generation than an allele having the opposite tendency. The selective advantage may be seen to lie through benefits conferred indifferently on a set of relatives each of which has a half chance of carrying the gene in question.

From this point of view it is also seen, however, that there is nothing special about the parent-offspring relationship except its close degree and a certain fundamental asymmetry. The full-sib relationship is just as close. If an individual carries a certain gene the expectation that a random sib will carry a replica of it is again one-half. Similarly, the half-sib relationship is equivalent to that of grandparent and grandchild with the expectation of replica genes, or genes "identical by descent" as they are usually called, standing at one quarter; and so on.

Although it does not seem to have received very detailed attention the possibility of the evolution of characters benefiting descendants more remote than immediate offspring has often been noticed. Opportunities for benefiting relatives, remote or not, in the same or an adjacent generation (i.e., relatives like cousins and nephews) must be much more common than opportunities for benefiting grandchildren and further descendants. As a first step toward a general theory that would take into account all kinds of relatives this paper will describe a model which is particularly adapted to deal with interactions between relatives of the same generation. The model includes the classical model for "nonoverlapping generations" as a special case. An excellent summary of the general properties of this classical model has been given by Kingman. It is quite beyond the author's power to give an equally extensive survey of the properties of the present model but certain approximate deterministic implications of biological interest will be pointed out.

As is already evident the essential idea which the model is going to use is quite simple. Thus although the following account is necessarily somewhat mathematical it is not surprising that eventually, allowing certain lapses from mathematical rigor, we are able to arrive at approximate principles which can also be expressed quite simply and in nonmathematical form. The most important principle, as it arises directly from the model, is outlined in the last section of this paper, but a fuller discussion together with some attempt to evaluate the theory as a whole in the light of biological evidence will be given in the sequel.

2. The Model

The model is restricted to the case of an organism which reproduces once and for all at the end of a fixed period. Survivorship and reproduction can both vary but it is only the consequent variations in their product, net reproduction, that are of concern here. All genotypic effects are conceived as increments and decrements to a basic unit of reproduction which, if possessed by all the individuals alike, would render the population both stationary and nonevolutionary. Thus the fitness $a^{\bullet}$ of an individual is treated as the sum of his basic unit, the effect δa of his personal genotype and the total e° of effects on him due to his neighbors which will depend on their genotypes:

$$a^{\bullet} = 1 + \delta a + e^{\circ}. \tag{1}$$

The index symbol $^{\bullet}$ in contrast to $^{\circ}$ will be used consistently to denote the inclusion of the personal effect δa in the aggregate in question. Thus equation (1) could be rewritten

$$a^{\bullet} = 1 + e^{\bullet}.$$

In equation (1), however, the symbol $^{\bullet}$ also serves to distinguish this neighbor modulated kind of fitness from the part of it

$$a = 1 + \delta a$$

which is equivalent to fitness in the classical sense of individual fitness.

The symbol δ preceding a letter will be used to indicate an effect or total of effects due to an individual treated as an addition to the basic unit, as typified in

$$a = 1 + \delta a.$$

The neighbors of an individual are considered to be affected differently according to their relationship with him.

Genetically two related persons differ from two unrelated members of the population in their tendency to carry replica genes which they have both inherited from the one or more ancestors they have in common. If we consider an autosomal locus, not subject to selection, in relative B with respect to the same locus in the other relative A, it is apparent that there are just three possible conditions of this locus in B, namely that both, one only, or neither of his genes are identical by descent with genes in A. We denote the respective probabilities of these conditions by c_2, c_1 and c_0. They are independent of the locus considered; and since

$$c_2 + c_1 + c_0 = 1,$$

the relationship is completely specified by giving any two of them. Li and Sacks have described methods of calculating these probabilities adequate for any relationship that does not involve inbreeding. The mean number of genes per locus i.b.d. (as from now on we abbreviate the phrase "identical by descent") with genes at the same locus in A for a hypothetical population of relatives like B is clearly $2c_2 + c_1$. One half of this number, $c_2 + \frac{1}{2}c_1$, may therefore be called the expected fraction of genes i.b.d. in a relative. It can be shown that it is equal to Sewall Wright's Coefficient of Relationship r (in a non-inbred population). The standard methods of calculating r without obtaining the complete distribution can be found in Kempthorne. Tables of

$$f = \tfrac{1}{2}r = \tfrac{1}{2}(c_2 + \tfrac{1}{2}c_1) \text{ and } F = c_2$$

for a large class of relationships can be found in Haldane and Jayakar.

Strictly, a more complicated metric of relationship taking

into account the parameters of selection is necessary for a locus undergoing selection, but the following account based on use of the above coefficients must give a good approximation to the truth when selection is slow and may be hoped to give some guidance even when it is not.

Consider now how the effects which an arbitrary individual distributes to the population can be summarized. For convenience and generality we will include at this stage certain effects (such as effects on parents' fitness) which must be zero under the restrictions of this particular model, and also others (such as effects on offspring) which although not necessarily zero we will not attempt to treat accurately in the subsequent analysis.

The effect of A on specified B can be a variate. In the present deterministic treatment, however, we are concerned only with the means of such variates. Thus the effect which we may write $(\delta a_{\text{father}})_{\text{A}}$ is really the expectation of the effect of A upon his father but for brevity we will refer to it as the effect on the father.

The full array of effects like $(\delta a_{\text{father}})_{\text{A}}$, $(\delta a_{\text{specified sister}})_{\text{A}}$, etc., we will denote

$$\{\delta a_{\text{rel.}}\}_{\text{A}}.$$

From this array we can construct the simpler array

$$\{\delta a_{r,c_2}\}_{\text{A}}$$

by adding together all effects to relatives who have the same values for the pair of coefficients (r, c_2). For example, the combined effect $\delta a_{\frac{1}{4},0}$ might contain effects actually occurring to grandparents, grandchildren, uncles, nephews and half-brothers. From what has been said above it is clear that as regards changes in autosomal gene-frequency by natural selection all the consequences of the full array are implied by this reduced array—at least, provided we ignore (a) the effect of previous generations of selection on the expected constitution of relatives, and (b) the one or more generations that must really occur before effects to children, nephews, grandchildren, etc., are manifested.

From this array we can construct a yet simpler array, or vector,

$$\{\delta a_r\}_A,$$

by adding together all effects with common r. Thus $\delta a_{\frac{1}{4}}$ would bring together effects to the above-mentioned set of relatives and effects to double-first cousins, for whom the pair of coefficients is $(\frac{1}{4}, \frac{1}{16})$.

Corresponding to the effect which A causes to B there will be an effect of similar type on A. This will either come from B himself or from a person who stands to A in the same relationship as A stands to B. Thus corresponding to an effect by A on his nephew there will be an effect on A by his uncle. The similarity between the effect which A dispenses and that which he receives is clearly an aspect of the problem of the correlation between relatives. Thus the term e° in equation (1) is not a constant for any given genotype of A since it will depend on the genotypes of neighbors and therefore on the gene-frequencies and the mating system.

Consider a single locus. Let the series of allelomorphs be $G_1, G_2, G_3, \ldots, G_n$, and their gene-frequencies $p_1, p_2, p_3, \ldots, p_n$. With the genotype G_iG_j associate the array $\{\delta a_{rel.}\}_{ij}$; within the limits of the above-mentioned approximations natural selection in the model is then defined.

If we were to follow the usual approach to the formulation of the progress due to natural selection in a generation, we should attempt to give formulae for the neighbor modulated fitnesses $\dot{a}_{ij}$. In order to formulate the expectation of that element of $\overset{\circ}{e}_{ij}$ which was due to the return effect of a relative B we would need to know the distribution of possible genotypes of B, and to obtain this we must use the double measure of B's relationship and the gene-frequencies just as in the problem of the correlation between relatives. Thus the formula for $\overset{\circ}{e}_{ij}$ will involve all the arrays $\{\delta a_{..,c_2}\}_{ij}$ and will be rather unwieldy (see Section 4).

An alternative approach, however, shows that the arrays $\{\delta a_r\}_{ij}$ are sufficient to define the selective effects. Every effect on reproduction which is due to A can be thought of as made up of two parts: an effect on the reproduction of genes i.b.d. with genes in A, and an effect on the reproduction of unrelated genes. Since the coefficient r measures the expected fraction of genes i.b.d. in a relative, for any particular degree of relationship this breakdown may be written quantitatively:

$$(\delta a_{\text{rel.}})_{\text{A}} = r(\delta a_{\text{rel.}})_{\text{A}} + (1 - r)(\delta a_{\text{rel.}})_{\text{A}}.$$

The total of effects on reproduction which are due to A may be treated similarly:

$$\sum_{\text{rel.}} (\delta a_{\text{rel.}})_{\text{A}} = \sum_{\text{rel.}} r(\delta a_{\text{rel.}})_{\text{A}} + \sum_{\text{rel.}} (1 - r)(\delta a_{\text{rel.}})_{\text{A}},$$

or

$$\sum_{r} (\delta a_{r})_{\text{A}} = \sum_{r} r(\delta a_{r})_{\text{A}} + \sum_{r} (1 - r)(\delta a_{r})_{\text{A}},$$

which we rewrite briefly as

$$\delta T_{\text{A}}^{\bullet} = \delta R_{\text{A}}^{\bullet} + \delta S_{\text{A}},$$

where $\delta R_{A}^{\bullet}$ is accordingly the total effect on genes i.b.d. in relatives of A, and δS_{A} is the total effect on their other genes. The reason for the omission of an index symbol from the last term is that here there is, in effect, no question of whether or not the self-effect is to be in the summation, for if it is included it has to be multiplied by zero. If index symbols were used we should have $\delta S_{A}^{\bullet} = \delta S_{A}^{\circ}$, whatever the subscript; it therefore seems more explicit to omit them throughout.

If, therefore, all effects are accounted to the individuals that cause them, of the total effect $\delta T_{ij}^{\bullet}$ due to an individual of genotype G_iG_j a part $\delta R_{ij}^{\bullet}$ will involve a specific contribution to the gene-pool by this genotype, while the remaining part δS_{ij} will involve an unspecific contribution consisting of genes in the ratio in which the gene-pool already possesses them. It is clear that it is the matrix of effects $\delta R_{ij}^{\bullet}$ which determines the direction of selection progress in gene-frequencies; δS_{ij} only influences its magnitude. In view of this importance of the $\delta R_{ij}^{\bullet}$ it is convenient to give some name to the concept with which they are associated.

In accordance with our convention let

$$R^{\bullet}_{ij} = 1 + \delta R^{\bullet}_{ij};$$

then $R^{\bullet}_{ij}$ will be called the *inclusive fitness*, $\delta R^{\bullet}_{ij}$ the *inclusive fitness effect* and δS_{ij} the *diluting effect*, of the genotype G_iG_j.

Let

$$T^{\bullet}_{ij} = 1 + \delta T^{\bullet}_{ij}.$$

So far our discussion is valid for nonrandom mating but from now on for simplicity we assume that it is random. Using a prime to distinguish the new gene-frequencies after one generation of selection we have

$$p'_i = \frac{\sum_j p_i p_j R^{\bullet}_{ij} + p_i \sum_{j,k} p_j p_k \delta S_{jk}}{\sum_{j,k} p_j p_k T^{\bullet}_{jk}} = p_i \frac{\sum_j p_j R^{\bullet}_{ij} + \sum_{j,k} p_j p_k \delta S_{jk}}{\sum_{j,k} p_j p_k T^{\bullet}_{jk}}.$$

The terms of this expression are clearly of the nature of averages over a part (genotypes containing G_i, homozygotes G_iG_i counted twice) and the whole of the existing set of genotypes in the population. Thus using a well-known subscript notation we may rewrite the equation term by term as

$$p'_i = p_i \frac{R^{\bullet}_{i.} + \delta S_{..}}{T^{\bullet}_{..}}$$

$$\therefore\ p'_i - p_i = \Delta p_i = \frac{p_i}{T^{\bullet}_{..}}(R^{\bullet}_{i.} + \delta S_{..} - T^{\bullet}_{..})$$

or

$$\Delta p_i = \frac{p_i}{R^{\bullet}_{..} + \delta S_{..}}(R^{\bullet}_{i.} - R^{\bullet}_{..}). \qquad (2)$$

This form clearly differentiates the roles of the $R^{\bullet}_{ij}$ and δS_{ij} in selective progress and shows the appropriateness of calling the latter diluting effects.

For comparison with the account of the classical case given by Moran, equation (2) may be put in the form

$$\Delta p_i = \frac{p_i}{T^{\bullet}_{..}}\left(\frac{1}{2}\frac{\partial R^{\bullet}_{..}}{\partial p_i} - R^{\bullet}_{..}\right)$$

where $\partial/\partial p_i$ denotes the usual partial derivative, written d/dp_i by Moran.

Whether the selective effect is reckoned by means of the $a_{ij}^{\bullet}$ or according to the method above, the denominator expression must take in all effects occurring during the generation. Hence $a_{..}^{\bullet} = T_{..}^{\bullet}$.

As might be expected from the greater generality of the present model the extension of the theorem of the increase of mean fitness presents certain difficulties. However, from the above equations it is clear that the quantity that will tend to maximize, if any, is $R_{..}^{\bullet}$, the mean inclusive fitness. The following brief discussion uses Kingman's approach.

The mean inclusive fitness in the succeeding generation is given by

$$R_{..}^{\bullet\prime} = \sum_{i,j} p_i' p_j' R_{ij}^{\bullet} = \frac{1}{T_{..}^{\bullet 2}} \sum_{i,j} p_i p_j R_{ij}^{\bullet} (R_{i.}^{\bullet} + \delta S_{..}) (R_{.j}^{\bullet} + \delta S_{..}).$$

$$\therefore \; R_{..}^{\bullet\prime} - R_{..}^{\bullet} = \Delta R_{..}^{\bullet} = \frac{1}{T_{..}^{\bullet 2}} \Big\{ \sum_{i,j} p_i p_j R_{ij}^{\bullet} R_{i.}^{\bullet} R_{.j}^{\bullet} +$$

$$+ \; 2\delta S_{..} \sum_{i,j} p_i p_j R_{ij}^{\bullet} R_{i.}^{\bullet} + R_{..}^{\bullet} \delta S_{..}^2 - R_{..}^{\bullet} T_{..}^{\bullet 2} \Big\}.$$

Substituting $R_{..}^{\bullet} + \delta S_{..}$ for $T_{..}^{\bullet}$ in the numerator expression, expanding and rearranging:

$$\Delta R^{\bullet} = \frac{1}{T_{..}^{\bullet 2}} \Big\{ \Big(\sum_{i,j} p_i p_j R_{ij}^{\bullet} R_{i.}^{\bullet} R_{.j}^{\bullet} - R_{..}^{\bullet 3} \Big) +$$

$$+ \; 2\delta S_{..} \Big(\sum_{i,j} p_i p_j R_{ij}^{\bullet} R_{i.}^{\bullet} - R_{..}^{\bullet 2} \Big) \Big\}.$$

We have $(\;\;) \geq 0$ in both cases. The first is the proven inequality of the classical model. The second follows from

$$\sum_{i,j} p_i p_j R_{ij}^{\bullet} R_{i.}^{\bullet} = \sum_i p_i R_{i.}^{\bullet 2} \geq \Big(\sum_i p_i R_{i.}^{\bullet} \Big)^2 = R_{..}^{\bullet 2}.$$

Thus a sufficient condition for $\Delta R_{..}^{\bullet} \geq 0$ is $\delta S_{..} \geq 0$. That

$\Delta R^{\bullet}_{..} \geq 0$ for positive dilution is almost obvious if we compare the actual selective changes with those which would occur if $\{R^{\bullet}_{ij}\}$ were the fitness matrix in the classical model.

It follows that $R^{\bullet}_{..}$ certainly maximizes (in the sense of reaching a local maximum of $R^{\bullet}_{..}$) if it never occurs in the course of selective changes that $\delta S_{..} < 0$. Thus $R^{\bullet}_{..}$ certainly maximizes if all $\delta S_{ij} \geq 0$ and therefore also if all $(\delta a_{\text{rel.}})_{ij} \geq 0$. It still does so even if some or all δa_{ij} are negative, for, as we have seen δS_{ij} is independent of δa_{ij}.

Here then we have discovered a quantity, inclusive fitness, which under the conditions of the model tends to maximize in much the same way that fitness tends to maximize in the simpler classical model. For an important class of genetic effects where the individual is supposed to dispense benefits to his neighbors, we have formally proved that the average inclusive fitness in the population will always increase. For cases where individuals may dispense harm to their neighbors we merely know, roughly speaking, that the change in gene frequency in each generation is aimed somewhere in the direction of a local maximum of average inclusive fitness, but may, for all the present analysis has told us, overshoot it in such a way as to produce a lower value.

As to the nature of inclusive fitness it may perhaps help to clarify the notion if we now give a slightly different verbal presentation. Inclusive fitness may be imagined as the personal fitness which an individual actually expresses in its production of adult offspring as it becomes after it has been first stripped and then augmented in a certain way. It is stripped of all components which can be considered as due to the individual's social environment, leaving the fitness which he would express if not exposed to any of the harms or benefits of that environment. This quantity is then augmented by certain fractions of the quantities of harm and benefit which the individual himself causes to the fitnesses of his neighbors. The fractions in question are simply the coefficients of relationship appropriate to the neighbors whom he affects: unity for clonal individuals, one-half for sibs, one-quarter for half-sibs, one-eighth for cousins, ...

and finally zero for all neighbors whose relationship can be considered negligibly small.

Actually, in the preceding mathematical account we were not concerned with the inclusive fitness of individuals as described here but rather with certain averages of them which we call the inclusive fitnesses of types. But the idea of the inclusive fitness of an individual is nevertheless a useful one. Just as in the sense of classical selection we may consider whether a given character expressed in an individual is adaptive in the sense of being in the interest of his personal fitness or not, so in the present sense of selection we may consider whether the character or trait of behavior is or is not adaptive in the sense of being in the interest of his inclusive fitness.

3. Three Special Cases

Equation (2) may be written

$$\Delta p_i = p_i \frac{\delta R_{i.}^{\bullet} - \delta R_{..}^{\bullet}}{1 + \delta T_{..}^{\bullet}}. \tag{3}$$

Now $\delta T_{ij}^{\bullet} = \sum_r (\delta a_r)_{ij}$ is the sum and $\delta R^{\bullet} = \sum_r r(\delta a_r)_{ij}$ is the first moment about $r = 0$ of the array of effects $\{\delta a_{\text{rel.}}\}_{ij}$ cause by the genotype G_iG_j; it appears that these two parameters are sufficient to fix the progress of the system under natural selection within our general approximation.

Let

$$r_{ij}^{\bullet} = \frac{\delta R_{ij}^{\bullet}}{\delta T_{ij}^{\bullet}}, \qquad (\delta T_{ij}^{\bullet} \neq 0); \tag{4}$$

and let

$$r_{ij}^{\circ} = \frac{\delta R_{ij}^{\circ}}{\delta T_{ij}^{\circ}}, \qquad (\delta T_{ij}^{\circ} \neq 0). \tag{5}$$

These quantities can be regarded as average relationships or as the first moments of reduced arrays, similar to the first moments of probability distributions.

We now consider three special cases which serve to bring out certain important features of selection in the model.

(a) The sums $\delta T^{\bullet}_{ij}$ differ between genotypes, the reduced first moment $r^{\bullet}$ being common to all. If all higher moments are equal between genotypes, that is, if all arrays are of the same "shape," this corresponds to the case where a stereotyped social action is performed with differing intensity or frequency according to genotype.

Whether or not this is so, we may, from equation (4), substitute $r^{\bullet}\delta T^{\bullet}_{ij}$ for $\delta R^{\bullet}_{ij}$ in equation (3) and have

$$\Delta p_i = p_i r^{\bullet} \frac{\delta T^{\bullet}_{i.} - \delta T^{\bullet}_{..}}{1 + \delta T^{\bullet}_{..}} .$$

Comparing this with the corresponding equation of the classical model,

$$\Delta p_i = p_i \frac{\delta a_{i.} - \delta a_{..}}{1 + \delta a_{..}} . \qquad (6)$$

we see that placing genotypic effects on a relative of degree $r^{\bullet}$ instead of reserving them for personal fitness results in a slowing of selection progress according to the fractional factor $r^{\bullet}$.

If, for example, the advantages conferred by a "classical" gene to its carriers are such that the gene spreads at a certain rate the present result tells us that in exactly similar circumstances another gene which conferred similar advantages to the sibs of the carriers would progress at exactly half this rate.

In trying to imagine a realistic situation to fit this sort of case some concern may be felt about the occasions where through the probabilistic nature of things the gene-carrier happens not to have a sib, or not to have one suitably placed to receive the benefit. Such possibilities and their frequencies of realization must, however, all be taken into account as the effects $(\delta a_{\text{sibs}})_{\text{A}}$, etc., are being evaluated for the model, very much as if in a classical case allowance were being made for some degree of failure of penetrance of a gene.

(b) The reduced first moments $r_{ij}^{\bullet}$ differ between genotypes, the sum $\delta T^{\bullet}$ being common to all. From equation (4), substituting $r_{ij}^{\bullet}\delta T^{\bullet}$ for $\delta R_{ij}^{\bullet}$ in equation (3) we have

$$\Delta p_i = p_i \frac{\delta T^{\bullet}}{T^{\bullet}}(r_{i.}^{\bullet} - r_{..}^{\bullet}).$$

But it is more interesting to assume δa is also common to all genotypes. If so it follows that we can replace $^{\bullet}$ by $^{\circ}$ in the numerator expression of equation (3). Then, from equation (5), substituting $r_{ij}^{\circ}\delta T^{\circ}$ for δR_{ij}°, we have

$$\Delta p_i = p_i \frac{\delta T^{\circ}}{T^{\bullet}}(r_{i.}^{\circ} - r_{..}^{\circ}).$$

Hence, if a giving-trait is in question (δT° positive), genes which restrict giving to the nearest relative ($r_{i.}^{\circ}$ greatest) tend to be favored; if a taking-trait (δT° negative), genes which cause taking from the most distant relatives tend to be favored.

If all higher reduced moments about $r = r_{ij}^{\circ}$ are equal between genotypes it is implied that the genotype merely determines whereabouts in the field of relationship that centers on an individual a stereotyped array of effects is placed.

With many natural populations it must happen that an individual forms the center of an actual local concentration of his relatives which is due to a general inability or disinclination of the organisms to move far from their places of birth. In such a population, which we may provisionally term "viscous," the present form of selection may apply fairly accurately to genes which affect vagrancy. It follows from the statements of the last paragraph but one that over a range of different species we would expect to find giving-traits commonest and most highly developed in the species with the most viscous populations whereas uninhibited competition should characterize species with the most freely mixing populations.

In the viscous population, however, the assumption of random mating is very unlikely to hold perfectly, so that these indications are of a rough qualitative nature only.

(c) $\delta T_{ij}^{\bullet} = 0$ for all genotypes.

$$\therefore \quad \delta T^{\circ}_{ij} = -\, \delta a_{ij}$$

for all genotypes, and from equation (5)

$$\delta R^{\circ}_{ij} = -\, \delta a_{ij} r^{\circ}_{ij}.$$

Then, from equation (3), we have

$$\Delta p_i = p_i(\delta R^{\bullet}_{i.} - \delta R^{\bullet}_{..}) = p_i\{(\delta a_{i.} + \delta R^{\circ}_{i.}) - (\delta a_{..} + \delta R^{\circ}_{..})\}$$
$$= p_i\{\cdot \delta a_{i.}(1 - r^{\circ}_{i.}) - \delta a_{..}(1 - r^{\circ}_{..})\}.$$

Such cases may be described as involving transfers of reproductive potential. They are especially relevant to competition, in which the individual can be considered as endeavoring to transfer prerequisites of survival and reproduction from his competitors to himself. In particular, if $r^{\circ}_{ij} = r^{\circ}$ for all genotypes we have

$$\Delta p_i = p_i(1 - r^{\circ})(\delta a_{i.} - \delta a_{..}).$$

Comparing this to the corresponding equation of the classical model [equation (6)] we see that there is a reduction in the rate of progress when transfers are from a relative.

It is relevant to note that Haldane in his first paper on the mathematical theory of selection pointed out the special circumstances of competition in the cases of mammalian embryos in a single uterus and of seeds both while still being nourished by a single parent plant and after their germination if they were not very thoroughly dispersed. He gave a numerical example of competition between sibs showing that the progress of gene-frequency would be slower than normal.

In such situations as this, however, where the population may be considered as subdivided into more or less standard-sized batches each of which is allotted a local standard-sized pool of reproductive potential (which in Haldane's case would consist almost entirely of prerequisites for pre-adult survival), there is, in addition to a small correcting term which we mention in the short general discussion of competition in the next section, an extra over-all slowing in selection progress. This may be thought of as due to the wasting of the powers of the more fit and the protection of the less fit when these types chance to occur positively assorted (beyond any mere effect of relationship) in a locality; its importance may be judged from the fact that it ranges from zero when the batches are indefinitely large to a halving of the rate of progress for competition in pairs.

4. Artificialities of the Model

When any of the effects is negative the restrictions laid upon the model hitherto do not preclude certain situations which are clearly impossible from the biological point of view. It is clearly absurd if for any possible set of gene-frequencies any $\overset{\bullet}{a}_{ij}$ turns out negative; and even if the magnitude of δa_{ij} is sufficient to make $\overset{\bullet}{a}_{ij}$ positive while $1 + e^{\circ}_{ij}$ is negative the situation is still highly artificial, since it implies the possibility of a sort of overdraft on the basic unit of an individual which has to be made good from his own takings. If we call this situation "improbable" we may specify two restrictions: a weaker, $e^{\circ}_{ij} > -1$, which precludes "improbable" situations; and a stronger, $\overset{\bullet}{e}_{ij} > -1$, which precludes even the impossible situations, both being required over the whole range of possible gene-frequencies as well as the whole range of genotypes.

As has been pointed out, a formula for $\overset{\bullet}{e}_{ij}$ can only be given if we have the arrays of effects according to a double coefficient of relationship. Choosing the double coefficient (c_2, c_1) such a formula is

$$\overset{\bullet}{e}_{ij} = \sum_{c_2, c_1}{}^{\bullet} \left[c_2 \operatorname{Dev}(\delta a_{c_2,c_1})_{ij} + \tfrac{1}{2}c_1\{\operatorname{Dev}(\delta a_{c_2,c_1})_{i.} + \operatorname{Dev}(\delta a_{c_2,c_1})_{.j}\}\right] + \delta T^{\circ}_{..}$$

where

$$\operatorname{Dev}(\delta a_{c_2,c_1})_{ij} = (\delta a_{c_2,c_1})_{ij} - (\delta a_{c_2,c_1})_{..} \text{ etc.}$$

Similarly

$$e^{\circ}_{ij} = \Sigma^{\circ}[''] + \delta T^{\circ}_{..},$$

the self-effect $(\delta a_{1,0})_{ij}$ being in this case omitted from the summations.

The following discussion is in terms of the stronger restriction but the argument holds also for the weaker; we need only replace $^{\bullet}$ by $^{\circ}$ throughout.

If there are no dominance deviations, i.e., if

$$(\delta a_{\text{rel.}})_{ij} = \tfrac{1}{2}\{(\delta a_{\text{rel.}})_{ii} + (\delta a_{\text{rel.}})_{jj}\} \text{ for all } ij \text{ and rel.,}$$

it follows that each ij deviation is the sum of the $i.$ and the $j.$ deviations. In this case we have

$$e^{\bullet}_{ij} = \Sigma\ {}^{\bullet}r\,\text{Dev}\,(\delta a_r)_{ij} + \delta T^{\bullet}_{..}.$$

Since we must have $e^{\bullet}_{..} = \delta T^{\bullet}_{..}$, it is obvious that some of the deviations must be negative.

Therefore $\delta T^{\bullet}_{..} > -1$ is a necessary condition for $e^{\bullet}_{ij} > -1$. This is, in fact, obvious when we consider that $\delta T^{\bullet}_{..} = -1$ would mean that the aggregate of individual taking was just sufficient to eat up all basic units exactly. Considering that the present use of the coefficients of relationships is only valid when selection is slow, there seems little point in attempting to derive mathematically sufficient conditions for the restriction to hold; intuitively however it would seem that if we exclude over- and underdominance it should be sufficient to have no homozygote with a net taking greater than unity.

Even if we could ignore the breakdown of our use of the coefficient of relationship it is clear enough that if $\delta T^{\bullet}_{..}$ approaches anywhere near -1 the model is highly artificial and implies a population in a state of catastrophic decline. This does not mean, of course, that mutations causing large selfish effects cannot receive positive selection; it means that their expression must moderate with increasing gene-frequency in a way that is inconsistent with our model. The "killer" trait of *Paramoecium* might be regarded as an example of a selfish trait with potentially large effects, but with its only partially genetic mode of inheritance and inevitable density dependence it obviously requires a selection model tailored to the case, and the same is doubtless true of most "social" traits which are as extreme as this.

Really the class of model situations with negative neighbor effects which are artificial according to a strict interpretation of the assumptions must be much wider than the class which we have chosen to call "improbable." The model assumes that the magnitude of an effect does not depend either on the genotype of the effectee or on his state with respect to the prerequisites of fitness at the time when the effect is caused. Where taking-traits are concerned it is just possible to imagine that this is true of some kinds of surreptitious theft but in general it is more reasonable to suppose that following some sort of an encounter the limited prerequisite is divided in the ratio of the competitive abilities. Provided competitive differentials are small however, the model will not be far from the truth; the correcting term that should be added to the expression for Δp_i can be shown to be small to the third order. With giving-traits it is more reasonable to suppose that if it is the nature of the prerequisite to be transferable the individual can give away whatever fraction of his own property that his instincts incline him to. The model was designed to illuminate altruistic behavior; the classes of selfish and competitive behavior which it can also usefully illuminate are more restricted, especially where selective differentials are potentially large.

For loci under selection the only relatives to which our metric of relationship is strictly applicable are ancestors. Thus the chance that an arbitrary parent carries a gene picked in an offspring is $\frac{1}{2}$, the chance that an arbitrary grandparent carries it is $\frac{1}{4}$, and so on. As regards descendants, it seems intuitively plausible that for a gene which is making steady progress in gene-frequency the true expectation of genes i.b.d. in a n-th generation descendant will exceed $\frac{1}{2}^n$, and similarly that for a gene that is steadily declining in frequency the reverse will hold. Since the path of genetic connection with a simple same-generation relative like a half-sib includes an "ascending part" and a "descending part" it is tempting to imagine that the ascending part can be treated with multipliers of exactly $\frac{1}{2}$

and the descending part by multipliers consistently more or less than $\frac{1}{2}$ according to which type of selection is in progress. However, a more rigorous attack on the problem shows that it is more difficult than the corresponding one for simple descendants, where the formulation of the factor which actually replaces $\frac{1}{2}$ is quite easy at least in the case of classical selection. and the author has so far failed to reach any definite general conclusions as to the nature and extent of the error in the foregoing account which his use of the ordinary coefficients of relationship has actually involved.

Finally, it must be pointed out that the model is not applicable to the selection of new mutations. Sibs might or might not carry the mutation depending on the point in the germ-line of the parent at which it had occurred, but for relatives in general a definite number of generations must pass before the coefficients give the true—or, under selection, the approximate—expectations of replicas. This point is favorable to the establishment of taking-traits and slightly against giving-traits. A mutation can, however, be expected to overcome any such slight initial barrier before it has recurred many times.

5. The Model Limits to the Evolution of Altruistic and Selfish Behavior

With classical selection a genotype may be regarded as positively selected if its fitness is above the average and as counter-selected if it is below. The environment usually forces the average fitness $a_{..}$ toward unity; thus for an arbitrary genotype the sign of δa_{ij} is an indication of the kind of selection. In the present case although it is $T^{\bullet}_{..}$ and not $R^{\bullet}_{..}$ that is forced toward unity, the analogous indication is given by the inclusive fitness effect $\delta R^{\bullet}_{ij}$, for the remaining part, the diluting effect δS_{ij}, of the total genotypic effect $\delta T^{\bullet}_{ij}$ has no influence on the kind of selection. In other words the kind of selection may be considered determined by whether the inclusive fitness of a genotype is above or below average.

We proceed, therefore, to consider certain elementary criteria which determine the sign of the inclusive fitness effect. The argument applies to any genotype and subscripts can be left out. Let

$$\delta T^{\circ} = k\delta \alpha. \tag{7}$$

According to the signs of δa and δT° we have four types of behavior as set out in the following diagram:

		Neighbors: gain; δT° + ve	Neighbors: lose; δT° − ve
Individual	gains; δa + ve	k + ve *Selected*	k − ve Selfish behavior ?
Individual	loses; δa − ve	k − ve Altruistic behavior ?	k + ve *Counter selected*

The classes for which k is negative are of the greatest interest, since for these it is less obvious what will happen under selection. Also, if we regard fitness as like a substance and tending to be conserved, which must be the case in so far as it depends on the possession of material prerequisites of survival and reproduction, k − ve is the more likely situation. Perfect conservation occurs if $k = -1$. Then $\delta T^{\bullet} = 0$ and $T^{\bullet} = 1$: the gene-pool maintains constant "volume" from generation to generation. This case has been discussed in Case (c) of section 3. In general the value of k indicates the nature of the departure from conservation. For instance, in the case of an altruistic action $|k|$ might be called the ratio of gain involved in the action: if its value is two, two units of fitness are received by neighbors for every one lost by an altruist. In the case of a selfish action, $|k|$ might be called the ratio of diminution: if its value is again two, two units of fitness are lost by neighbors for one unit gained by the taker.

The alarm call of a bird probably involves a small extra risk to the individual making it by rendering it more noticeable to the approaching predator but the consequent reduction of risk to a nearby bird previously unaware of danger must be much greater.* We need not discuss here just how risks are to be reckoned in terms of fitness: for the present illustration it is reasonable to guess that for the generality of alarm calls k is negative but $|k| > 1$. How large must $|k|$ be for the benefit to others to outweigh the risk to self in terms of inclusive fitness?

$$
\begin{aligned}
\delta R^{\bullet} &= \delta R^{\circ} + \delta a \\
&= r^{\circ}\delta T^{\circ} + \delta a && \text{from (5)} \\
&= \delta a(kr^{\circ} + 1) && \text{from (7).}
\end{aligned}
$$

Thus of actions which are detrimental to individual fitness (δa – ve) only those for which $-k > 1/r^{\circ}$ will be beneficial to inclusive fitness ($\delta R^{\bullet}$ + ve).

This means that for a hereditary tendency to perform an action of this kind to evolve the benefit to a sib must average at least twice the loss to the individual, the benefit to a half-sib must be at least four times the loss, to a cousin eight times and so on. To express the matter more vividly, in the world of our model organisms, whose behavior is determined strictly by genotype, we expect to find that no one is prepared to sacrifice his life for any single person but that everyone will sacrifice it when he can thereby save more than two brothers, or four half-brothers, or eight first cousins.... Although according to the model a tendency to simple altruistic transfers

* The alarm call often warns more than one nearby bird of course—hundreds in the case of a flock—but since the predator would hardly succeed in surprising more than one in any case the total number warned must be comparatively unimportant.

($k = -1$) will never be evolved by natural selection, such a tendency would, in fact, receive zero counter-selection when it concerned transfers between clonal individuals. Conversely selfish transfers are always selected except when from clonal individuals.

As regards selfish traits in general (δa + ve, k − ve) the condition for a benefit to inclusive fitness is $-k < 1/r°$. Behavior that involves taking too much from close relatives will not evolve. In the model world of genetically controlled behavior we expect to find that sibs deprive one another of reproductive prerequisites provided they can themselves make use of at least one half of what they take; individuals deprive half-sibs of four units of reproductive potential if they can get personal use of at least one of them; and so on. Clearly from a gene's point of view it is worthwhile to deprive a large number of distant relatives in order to extract a small reproductive advantage.

3: *The Genetical Evolution of Social Behavior. II*

W. D. HAMILTON

Grounds for thinking that the model described in the previous paper can be used to support general biological principles of social evolution are briefly discussed.

Two principles are presented, the first concerning the evolution of social behavior in general and the second the evolution of social discrimination. Some tentative evidence is given.

More general application of the theory in biology is then discussed, particular attention being given to cases where the indicated interpretation differs from previous views and to cases which appear anomalous. A hypothesis is outlined concerning social evolution in the Hymenoptera; but the evidence that at present exists is found somewhat contrary on certain points. Other subjects considered include warning behavior, the evolution of distasteful properties in insects, clones of cells and clones of zooids as contrasted with other types of colonies, the confinement of parental care to true offspring in birds and insects, fights, the behavior of parasitoid insect larvae within a host, parental care in connection with monogyny and monandry and multi-ovulate ovaries in plants in connection with wind and insect pollination.

1. Introduction

In the previous paper a genetical mathematical model was used to deduce a principle concerning the evolution of social behavior which, if true generally, may be of considerable importance in biology. It has now to be considered whether there is any logical justification for the extension of this principle beyond the model case of nonoverlapping generations, and,

From the *Journal of Theoretical Biology*, 7 (1964), 17–51. Editor's note: The notes and addendum were written by the author for this volume.

if so, whether there is evidence that it does work effectively in nature.

In brief outline, the theory points out that for a gene to receive positive selection it is not necessarily enough that it should increase the fitness of its bearer above the average if this tends to be done at the heavy expense of related individuals, because relatives, on account of their common ancestry, tend to carry replicas of the same gene; and conversely that a gene may receive positive selection even though disadvantageous to its bearers if it causes them to confer sufficiently large advantages on relatives. Relationship alone never gives grounds for *certainty* that a person carries a gene which a relative is known to carry except when the relationship is "clonal" or "mitotic" (e.g., the two are monozygotic twins)—and even then, strictly, the possibility of an intervening mutation should be admitted. In general, it has been shown that Wright's Coefficient of Relationship r approximates closely to the chance that a replica will be carried. Thus if an altruistic trait is in question more than $1/r$ units of reproductive potential or "fitness" must be endowed on a relative of degree r for every one unit lost by the altruist if the population is to gain on average more replicas than it loses. Similarly, if a selfish trait is in question, the individual must receive and use at least a fraction r of the quantity of "fitness" deprived from his relative if the causative gene is to be selected....

2. The Grounds for Generalization

It is clear that in outline this type of argument is not restricted to the case of nonoverlapping generations nor to the state of panmixia on which we have been able to base a fairly precise analysis. The idea of the regression, or "probabilistic dilution," of "identical" genes in relatives further and further removed applies to all organisms performing sexual reproduction, whether or not their generations overlap and whether or not the relatives considered belong to the same generations.

However perhaps we should not feel entirely confident

about generalizing our principle until a more comprehensive mathematical argument, with inclusive fitness more widely defined, has been worked out. But even from this point of view there does seem to be good reason for thinking that it can be generalized—reason about as good, at least, as that which is supposed to give foundation to certain principles of the classical theory.

Roughly speaking the classical mathematical theory has developed two parallel branches which lie to either side of the great range of reproductive schedules which organisms actually do manifest. One is applicable to once-and-for-all reproduction; and this form is actually exhibited by many organisms, notably those with annual life-cycles. The other is applicable to "continuous" reproduction. This involves a type of reproductive process which is strictly impossible for any organism to practice, but which for analytic purposes should be approximated quite closely by certain species, for example, some perennial plants. Our model is a generalization in the former branch and there seems little reason to doubt that it can be matched by a similar model in the latter.

Even in the classical theory itself difficulties still face generalization between the two branches, and yet their continuance does not seem to cause much worry. For instance there does not seem to be any comprehensive definition of fitness. And, perhaps in consequence of this lack, it rather appears that Fisher's Fundamental Theorem of Natural Selection has yet to be put in a form which is really as general as Fisher's original statement purports to be. On the other hand, the clarity of Fisher's statement must surely, for general usefulness, have far outweighed its defects in rigor.

3. Valuation of the Welfare of Relatives

Altogether then it would seem that generalization would not be too foolhardy. In the hope that it may provide a useful summary we therefore hazard the following generalized unrigorous statement of the main principle that has emerged from the model.

The social behavior of a species evolves in such a way that in each distinct behavior-evoking situation the individual will seem to value his neighbors' fitness against his own according to the coefficients of relationship appropriate to that situation.

The aspect of this principle which concerns altruism seems to have been realized by Haldane as is shown in some comments on whether a genetical trait causing a person to risk his life to save a drowning child could evolve or not. His argument, though not entirely explicit and apparently restricted to rare genes, is essentially the same as that which we have outlined for altruism in the Introduction.

Haldane does not discuss the question which his remarks raise of whether a gene lost in an adult is worth more or less than a gene lost in a child. However, this touches an aspect of the biological accounting of risks which together with the whole problem of the altruism involved in parental care is best reserved for separate discussion.

The principle was also foreshadowed much earlier in Fisher's discussion of the evolution of distastefulness in insects. That this phenomenon presents a difficulty, namely an apparent absence of positive selection, is obvious as soon as we reject the pseudo-explanations based on the "benefit to the species," and the problem is of considerable importance as distastefulness, construed in a wide sense, is the basis not only of warning coloration but of both Batesian and Mullerian mimicry. The difficulty of explaining the evolution of warning coloration itself is perhaps even more acute here; *a priori* we would expect that at every stage it would be the new ultra-conspicuous mutants that suffered the first attacks of inexperienced predators. Fisher suggested a benefit to the nearby siblings of the distasteful, or distasteful and conspicuous insect, and gave some suggestive evidence that these characters are correlated with gregariousness of the larvae. He remarked that "the selective potency of the avoidance of brothers will of course be only half as great as if the individual itself were protected; against this is to be set the fact that it applies to the whole of a possibly numerous brood." He doubtlessly realized that further selective benefit would occur through more distant relatives but probably considered it negligible. He realized the logical affinity of this problem with that of the evolution of altruistic behavior, and he invokes the same kind of selection in his attempt to explain the evolution of the heroic ideal in barbaric human societies.

Another attempt to elucidate the genetical natural selection of altruistic behavior occurring within a sibship was published by Williams and Williams in 1957. Although their conclusions are doubtlessly correct the particular form of analysis they adopted seems to have failed to bring out the crucial role of the two-fold factor in this case.

A predator would have to taste the distasteful insect before it could learn to avoid the nearby relatives. Thus despite the toughness and resilience which is supposed to characterize such insects (qualities which the classical selectionists may have been tempted to exaggerate), the common detriment to the "altruist" must be high and the ratio of gain to loss (k) correspondingly low. The risks involved in giving a warning signal, as between birds, must be much less so that in this case, as indicated in the previous paper, it is more credible that the condition

$$k > \frac{1}{\bar{r}}$$

is fulfilled even when cases of the parents warning their young and the young each other up to the time of their dispersal are left out of account. The average relationship within a rabbit-warren is probably quite sufficient to account for their "thumping" habit. Ringing experiments on birds indicate that even adult territorial neighbors must often be much closer relatives than their powers of flight would lead us to expect, a fact that may be of significance for the interpretation of the wider comity of bird behavior.

The phenomena of mutual preening and grooming may be

explained similarly. The mild effort required must stand for a diminution of fitness quite minute compared to the advantage of being cleansed and cleared of ectoparasites on parts of the body which the individual cannot deal with himself. Thus the degree of relationship within the flocks of birds, troupes of monkeys and so on where such mutual help occurs need not be very high before the condition for an advantage to inclusive fitness is fulfilled; and for grooming within actual families, of monkeys for instance, it is quite obviously fulfilled.

An animal whose reproduction is definitely finished cannot cause any further self-effects. Except for the continuing or pleiotropic effects of genes which are established through an advantage conferred earlier in the life-history, the behavior of a postreproductive animal may be expected to be entirely altruistic, the smallest degree of relationship with the average neighbor being sufficient to favor the selection of a giving trait. Blest has recently shown that the postreproductive behavior of certain saturnid moths is indeed adaptive in this way. His argument may be summarized in the present terminology as follows. With a species using cryptic resemblance for its protection the very existence of neighbors involves a danger to the individual since the discovery of one by a predator will be a step in teaching it to recognize the crypsis. With an aposematic species on the other hand, the existence of neighbors is an asset since they may well serve to teach an inexperienced predator the warning pattern. Thus with the cryptic moth it is altruistic to die immediately after reproduction, whereas with the warningly-colored moth it is altruistic to continue to live at least through the period during which other moths may not have finished mating and egg-laying. Blest finds that the postreproductive life-spans of the moths he studied are modified in the expected manner, and that the cryptic species even show behavior which might be interpreted as an attempt to destroy their cryptic pattern and to use up in random flight activity the remainder of their vital reserves. The selective forces operating on the postreproductive life-span are doubtless generally weak; they will be strongest when the average relationship of neighbors is highest, which will be in the most viscous populations. It would be interesting to know how behavior affecting gene-dispersion correlates with the degree of the effects which Blest has observed.

4. Discrimination in Social Situations

Special case (b) of the previous paper has shown explicitly that a certain social action cannot in itself be described as harmful or beneficial to inclusive fitness; this depends on the relationship of the affected individuals. The selective advantage of genes which make behavior conditional in the right sense on the discrimination of factors which correlate with the relationship of the individual concerned is therefore obvious. It may be, for instance, that in respect of a certain social action performed toward neighbors indiscriminately, an individual is only just breaking even in terms of inclusive fitness. If he could learn to recognize those of his neighbors who really were close relatives and could devote his beneficial action to them alone an advantage to inclusive fitness would at once appear. Thus a mutation causing such discriminatory behavior itself benefits inclusive fitness and would be selected. In fact, the individual may not need to perform any discrimination so sophisticated as we suggest here; a difference in the generosity of his behavior according to whether the situations evoking it were encountered near to, or far from, his own home might occasion an advantage of a similar kind.

Although this type of advantage is itself restricted to social situations, it can be compared to the general advantages associated with making responses conditional on the factors which are the most reliable indicators of future events, an advantage which must for instance have been the basis for the evolution of the seed's ability to germinate only when conditions (warmth, moisture, previous freezing, etc.) give real promise for the future survival and growth of the seedling.

Whether the trend implied could ever spread very far may be doubted. All kinds of evolutionary changes in behavior, especially those subject to the powerful forces of individual advantage, are liable to disrupt any *ad hoc* system of discrimination. This is most true, however, for discrimination in the range of distant relationships where the potential gains are least. The selective advantage when a benefit comes to be given to sibs only instead of to sibs and half-sibs indifferently is more than four times the advantage when a benefit of the same magnitude is given to cousins only instead of to cousins and half-cousins indifferently.

Nevertheless, if any correlate of relationship is very persistent, long-continued weak selection could lead to the evolution of a discrimination based on it even in the range of distant relationships. One possible factor of this kind in species with viscous populations, and one whose persistence depends only on the viscosity and therefore may well be considerably older than the species in question, is familiarity of appearance. For in a viscous population the organisms of a particular neighborhood, being relatives, must tend to look alike and an individual which used the restrained symbolic forms of aggressive behavior only toward familiar-looking rivals would be effecting a discrimination advantageous to inclusive fitness.

In accordance with the hypothesis that such discriminations exist it should turn out that in a species of resident bird, strongly territorial and minimally vagrant, the conflicts which proved least readily resolved by ritual behavior and in which consequent fighting was fiercest were between the rivals that had the most noticeable differences in plumage and song. Whether much evidence of this nature exists I do not know. The rather uncommon cases of interspecific territory systems in birds, as recently reviewed by Wynne-Edwards, seem to be contrary. If differences between interspecific and conspecific encounters were noticed by the original observers they are not mentioned by Wynne-Edwards; and in any case, the very existence of these situations, taken at face value and assumed to be stable and of long standing, is as contrary to the present theory as it is to Gause's principle. Likewise, the positive indications I can bring forward are rather few and feeble. Tinbergen has observed a hostile reaction by Herring Gulls toward members of their colony forced to behave abnormally (caught in a net) and states that a similar phenomenon is sometimes observed with other social species. Personal observations on colonies of the wasps *Polistes canadensis* and *P. versicolor* have shown a very strong hostility when a wasp, taken off a nest, is returned to it in a wet and bedraggled condition. This type of reaction after a member of the colony has been much handled seems to be quite common in the social insects. It is perhaps specifically aroused by certain acquired odors, or these combined with the odor of venom. That bird-ringers, who would surely have noticed any social stigma that fell upon birds carrying their often very conspicuous rings, usually report that the rings were no apparent inconvenience to the birds is a counter-indication whose force is slightly reduced by the fact that in passerines and most other common birds the legs are unimportant in social communication. It is similarly fortunate for the insect ethologist that spots of fresh oil-paint by themselves on bees and wasps seem to provoke very little reaction. Butterflies of the family *Lycaenidae*, especially males, are often to be seen jostling one another in the air, sometimes in groups of more than two. The function of this behavior is obscure; the species do not seem to be at all strongly territorial. According to Ford lepidopterists find that a bunch of jostling butterflies is rather apt to contain an unusual variety.

With the higher animals we may perhaps appeal to evidence of discrimination based on familiarity of a more intimate kind. Animals capable of forming a social hierarchy presumably have some ability to recognize one another as individuals, and with this present it is not necessary for the discrimination to be on the basis of "racialistic" differences of appearance, voice, or smell. An individual might look extremely like certain members of a group and lie within the group's range of variation in every one of his perceptible characters and yet still be known for a stranger. Speaking from a wide knowledge of just such social animals, Wynne-Edwards refers to "the widespread practice of attacking and persecuting strangers and relegating newcomers to the lowest social rank" and gives several references. The antagonistic nature of this discrimination is of course just what we expect.

As might be expected the evidence in the cases of closest relationship is much more impressive. Tinbergen investigated the ability of Herring Gulls to recognize their own chicks by observing the reaction to strange chicks placed among them. He found that during the first two or three days after hatching strange chicks are accepted, but by the end of the first week they are driven away. Herring Gulls will sometimes form the habit of feeding on the live chicks as well as on the eggs in their own breeding colony when they can catch them unattended, but Tinbergen records no case where an intruded chick was killed although this probably sometimes happens; the hostile behavior he observed was half-hearted at first but became more definite as the age of the gull's own brood advanced. During the days which follow hatching, the chicks become progressively more mobile and the chance that they will wander into neighboring nest-territories must increase. Therefore it seems a reasonable hypothesis that the ability to discriminate "own young" advances in step with the chance that without such discrimination strange chicks would be fostered and the benefits of parental care wasted on unrelated genes. Supporting this hypothesis are the findings quoted by Tinbergen of Watson and Lashley on two tropical species of tern: "The Noddies nesting in trees do not recognize their young at any age, whereas the ground-nesting Sooties are very similar to Herring Gulls in that they learn to recognize their own young in the course of four days." House Sparrows will accept strange young of the right age placed in the nest but after the nestlings have flown "they will not, in normal circumstances, feed any but their own young." Not all observations are as satisfactory for the theory as these however: we may mention the positive passion for fostering said to be shown by Emperor Penguins that have lost their own chick. This and some other similar anomalies will be briefly discussed in the last section.

Tinbergen showed that Herring Gulls discriminate eggs

even less than chicks, the crudest egg-substitutes being sufficient to release brooding behavior providing certain attributes of shape and color are present. This is what we would expect in view of the fact that eggs do not stray at all. It is in striking contrast with the degree of egg-discrimination which is shown by species of birds subject to cuckoo parasitism.

The theoretical principle which these observations seem largely to support is supplementary to the previous principle and we may summarize it in a similar statement.

> *The situations which a species discriminates in its social behavior tend to evolve and multiply in such a way that the coefficients of relationship involved in each situation become more nearly determinate.*

In situations where relationship is not variable, as for example between the nestlings in an arboreal nest, there still remains a discrimination which, if it could be made could greatly benefit inclusive fitness. This is the discrimination of those individuals which do carry one or both of the behavior-causing genes from those which do not. Such an ability lies outside the conditions postulated in the previous paper but the extended meaning of inclusive fitness is obvious enough. That genes could cause the perception of the presence of like genes in other individuals may sound improbable; at simplest we need to postulate something like a supergene affecting (a) some perceptible feature of the organism, (b) the perception of that feature, and (c) the social response consequent upon what was perceived. However, exactly the same *a priori* objections might be made to the evolution of assortative mating which manifestly has evolved, probably many times independently and despite its obscure advantages.

If some sort of attraction between likes for purposes of cooperation can occur the limits to the evolution of altruism expressed by our first principle would be very greatly extended, although it should still never happen that one individual would value another more highly than itself, fitness for fitness. And if an individual can be attracted toward likes when it has positive effects—benefits—to dispense, it can presumably be attracted the other way, toward unlikes, when it has negative effects to dispense (i.e., when circumstances arise which demand combat, suggest robbery, and so on).

5. Genetical Relationship in Colonies

In this section we discuss a small selection of the biological problems relating to life in colonies, choosing particularly those which the theory we have developed is able to illuminate in a simple and novel manner and those concerning which discussions in the existing literature are often unsatisfactory.

Clones

According to considerations advanced so far the coefficient of relationship between all members of a clone should be unity. If this is so our theory predicts for clones a complete absence of any form of competition which is not to the over-all advantage and also the highest degree of mutual altruism. This is borne out well enough by the behavior of the clones which make up the bodies of multicellular organisms. However, when we consider populations of free-living asexual organisms there appears to be a discrepancy in that competitive adaptation is hardly less conspicuous than it is for most wholly sexual populations and altruism, if it exists, is not easily detected. To account for this discrepancy three points may be made.

In the first place it may be doubted how many apparently asexual populations are really as they seem. Repeated discoveries of sexual or recombinative processes in species formerly thought to possess none may cause a suspicion that pure clonal populations of any considerable size are uncommon; and taking into account the well-known generalization that asexual reproduction tends to give place to sexual with the onset of adverse conditions, it may be argued that fully competitive (i.e., stationary or declining) pure clonal populations must be less common still. In a mixed sexual-asexual population the levels of competition and altruism should, neglecting mutation, be appropriate to the average relationship.

Second, as regards the appearance of competitive adaptations, we may repeat what was noted in the previous paper, namely that to the new mutant all individuals have zero relationship (for the locus in question); any selfish mutation must therefore have an immediate advantage and its progress will be merely slowed down, not completely arrested, by the self-destruction it comes to work in the later stages of its spread.

Third, as regards the absence of cooperation and altruism, we may note an adjustment to the metric of relationship which we have so far found it convenient to neglect but which will have a slight effect in reducing the relationship between individuals in a clonal population. This again involves mutation. Each step in the path of mitotic connection between two asexual organisms correponds to a constant chance of mutation (m). The chance that a mutation does not occur $(1-m)$ can be multiplied along these paths just as is the factor 1/2 along paths of meiotic connection in the ordinary calculation of r, and the grand product is likewise the expectation of replica genes in the relative. The number of generations for a given value of r to be reached is approximated by the formula

$$\frac{1}{2m}\log_e\frac{1}{r}.$$

This would apply to the minimum relationship but it is that borne to an individual by half the population and the average relationship is very close to it. With normal mutation rates the decrease in relationship will be slow. Thus if $m = 10^{-5}$ the number of generations for asexual descendants of a common ancestor to become as widely related as full-sibs or the gametes of a single sexual individual is about 39,660. A bacterium with continuously favorable growth conditions so that it divided once every 20 minutes would take $1\frac{1}{2}$ years to run through this many generations while a unicellular green alga such as *Chlorella*, dividing once every 15 hours, would take 68 years.

However, taking all three points together and especially considering the fact that a population will normally be started by many sexually produced spores, our apparent discrepancy is largely removed. Such obvious differences in cooperation and altruism as are apparent between a "*colony*" of *Volvox* and a *population* of *Chlamydomonas*, or, to present the contrast another way, within and between colonies of *Volvox*, are at least plausibly accounted for. The cooperation of the cells in the *Volvox* colony, or coenobium as it is perhaps better called, can be regarded as due to the closeness of their relationship, a mere 14 cell generations being necessary to produce the 10,000 or so cells concerned (*V. globulina*).

Thus the classical "evolutionary" series in the *Chlorophyceae*, starting with temporary cohesion of mitotic daughter-cells of a free-living unicellular form like *Chlamydomonas* and ending with forms with a large and highly differentiated soma, is well in accord with our theory.

Fusion of Individuals or Clones

If on the contrary such integrated colonies were found to be formed by the coming together of random members of the population or even by the cohesion of meiotic daughter-cells, there would be some cause for surprise, especially if a soma were formed without any sign of discord among the cells.

Something like this has in fact been noted in the *Rhodophyceae*. The sporelings developed from either carpospores or tetraspores of *Gracillaria verrucosa* were found to fuse readily when they grew into contact. Jones suggested that the compound sporelings so formed might have an advantage over solitary ones in nature in being less likely to be smothered by sand in the littoral situations in which they grow since he had observed that they sent up fronds sooner and more strongly; but about four out of five of the component sporelings must nevertheless have been total losers by the arrangement to judge by the numbers of fronds sent up. Jones does not state whether the spores in question were from a single parent thallus, but he states that he has seen young plants resembling his compound sporelings in the wild.

Fusion of plasmodia is known in the Myxomycetes and Acrasiales. But again, if the cultures in which this has been observed were made up from spores taken from a single sporulating plasmodium, as seems quite likely, the congregating cells or fusing plasmodia cannot be regarded as unrelated, and they could be segregants which happened to have received like combinations of the incompatibility genes normally effective in preventing fusion.

Knight-Jones and Moyse give an interesting summary of the known facts concerning fusion in marine colonial animals (including reference to the above-mentioned case of *Gracillaria*). It seems that fusion of adjacent colonies does sometimes occur naturally in sponges and corals when contact is made in the early stages of growth; but old colonies tend to develop a line of demarcation where they meet and the same is true of the Bryozoa and the colonial ascidians, fusion even in the early stages being unknown in these groups.

The theoretical considerations which the present theory would apply to the cases of the three preceding paragraphs may be gathered from the discussions that will be given in the next section concerning fighting and cooperation. In general, it is fair to state as a matter of fact that the sexually produced individuals of a species do not, and usually will not, fuse with one another. Of course from such a statement, a large exception must be made for the fusion of haploids in the normal sexual cycle; but here it will be noted that except in respect of certain unusual types of chromosomes the discipline of the meiotic process must generally assure equal reproductive expectations for the two cooperating genomes.

Knight-Jones and Moyse emphasize the contrast between the mutual behavior of zooids of a single colony and that occurring between the members of the dense clusters that arise from the gregarious settling of larvae: "Such systems are strikingly more economical than is a barnacle population, in that the crowded and smothered barnacles die wastefully, but unsuccessful zooids are resorbed and their materials presumably transferred to help growth elsewhere." According to the present view, clonal colonies of zooids are things of a very distinct kind from colonies of sexually produced organisms such as oysters or barnacles, and the cooperation of zooid individuals, which comes to reach such remarkable complexity in some of the pelagic Siphonophora, should in itself cause no surprise.

Colonies of Social Insects

The colonies of the social insects are remarkable in having true genetic diversity in the cooperating individuals.

Caution is necessary in applying the present theory to Hymenoptera because of course their system of sex-determination gives their population genetics a peculiar pattern. But there seems to be no reason to doubt that the concept of inclusive fitness is still valid.

a. *A hypothesis concerning the social tendencies of the Hymenoptera.* Using this concept it soon becomes evident that family relationships in Hymenoptera are potentially very favorable to the evolution of reproductive altruism.

If a female is fertilized by only one male all the sperm she receives is genetically identical. Thus, although the relationship of a mother to her daughters has the normal value of $\frac{1}{2}$, the relationship between daughters is $\frac{3}{4}$. Consider a species where the female consecutively provisions and oviposits in cell after cell so that she is still at work when the first of her female offspring ecloses, leaves the nest and mates. Our principle tells us that even if this new adult had a nest ready constructed and vacant for her use she would prefer, other things being equal, returning to her mother's and provisioning a cell for the rearing of an extra sister to provisioning a cell for a daughter of her own. From this point of view therefore it seems not surprising that social life appears to have had several independent origins in this group of insects or that certain divisions of it, represented mainly by solitary species which do more or less approximate the model situation (e.g., most halictine bees), do show sporadic tendencies toward the matrifilial colony.

It may seem that if worker instincts were so favored colony reproduction could never be achieved at all. However, this problem is more apparent than real. As soon as either the architectural difficulties of further adding to the nest, or a local shortage of food, or some other cumulative hindrance, makes the adding of a further bio-unit to the colony $1\frac{1}{2}$ times more difficult than the creating of the first bio-unit of a new colony the females should tend to go off to found new colonies. Of course, in a more advanced state with differentiated workers, the existing workers would be expected to connive at the change-over to the production queens, which is, so to speak, the final object of their altruism. That in actual species the change-over anticipates the onset of adverse conditions is not surprising since they must be to a large extent predictable. In Britain where winter sets the natural termination the vespine wasps round off their colony growth at about the time one would expect but some bumblebees begin rather surprisingly early. If climatic termination were not in question and queen-production tended to come a little late so that the worker population had already risen above the number that could work efficiently on the nest workers might best serve their inclusive fitness by going off with the dispersing queens, despite the fact that in this case the special high relationship of workers to the progeny of the queen no longer holds. Descriptively this is roughly what happens in the meliponine bees and, apart from the serious complication of the swarms having many queens each, it seems to be what happens in the polybiine wasps. In *Apis*, as is well known, it is the old queen who goes off with some of her daughters, *leaving* a young queen together with sister workers. This oddity cannot be so easily derived in the imagination from semi-social antecedents in colony re-

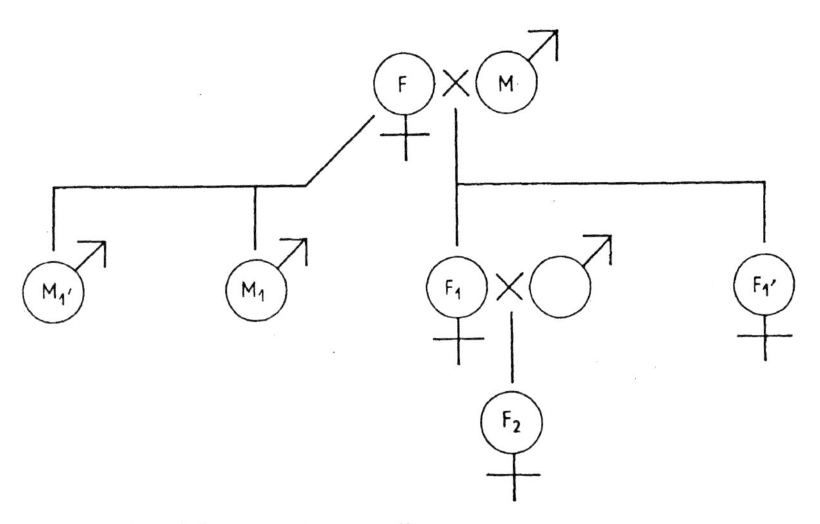

FIGURE 1: A hymenopteran pedigree.

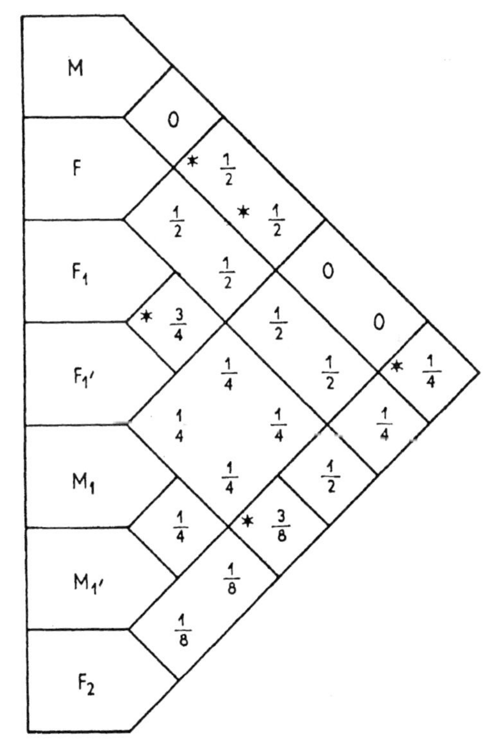

FIGURE 2: Coefficients of relationship for the pedigree of Figure 1. Asterisks indicate the coefficients that would diminish in cases of polyandrous insemination assuming the fathership of particular offspring to be unknown.

production (it could come more readily from the habit of the whole colony absconding under adverse conditions) and like other peculiar features in honeybees it hints at a long and complicated background of social evolution. Of course as attempts to represent the actual course of evolution and its forms of selection the above outlines are in any case thoroughly naive; they are merely intended to *illustrate* certain possible courses which would accord with our principles.

The idea that the male-haploid system of sex-determination contributes to the peculiar tendency of the Hymenoptera toward social evolution is somewhat strengthened by considering other relationships which may be relevant.

Figure 1 shows a hymenopteran pedigree and Figure 2 shows the coefficients of relationship between the individuals lettered on the pedigree.

The relationships concerning males are worked out by assuming each male to carry a "cipher" gene to make up his diploid pair, one "cipher" never being considered identical by descent with another.* For all male relationships we then have

$$r = \tfrac{1}{2}c_1$$

where c_1 is the chance that the two have a replica each. The convenience of this procedure, which is arbitrary in the sense that some other value for the fundamental mother-son and father-daughter link would have given an equally coherent system, is that it results in male and female offspring having equal relationships to their mother which matches with the fact that when the sex-ratio is in its equilibrium condition individuals of opposite sex have equal reproductive values (see Bodmer and Edwards in this volume).

The relationships whose values are affected by polyandrous insemination of the female are indicated in Figure 2 by asterisks. It will be seen that among those unaffected, because fertilization

* The rule given here for coefficients involving males is based on an error and must be discarded. In Figure 2 and p. 63, line 18, the brother-brother coefficient should be $\tfrac{1}{2}$. Values for heterosexual coefficients require qualification as explained in the addendum (pp. 87f.). Points made in discussion, however, are hardly affected.

is not involved, are the relationships of a female to her son, $r = \frac{1}{2}$, and to her brother, $r = \frac{1}{4}$. According to our theory these values indicate that workers should be much less inclined to give up their male-producing in favor of the queen's than they are to give up their female-producing in favor of a singly-mated queen. Laying by workers is known to occur in each of the main social groups, bees, wasps and ants. The extent to which the practice occurs in normal colonies remains largely obscure; but in some species it is so prevalent that observers have been led to suggest that all the male members of the population are produced in this way. In fairness, however, rather than emphasize this apparently detailed fit of our hypothesis, it should be pointed out that male-egg production by workers is in any case the simplest possible manifestation of an incipient selfish tendency since it does not require the complicated preliminary of mating.

Males are related to their brothers as well as to their sisters with $r = \frac{1}{4}$; their relationship to their daughters is $\frac{1}{2}$. Hence the favorable situation for the evolution of worker-like instincts cannot ever apply to males, and in conformity with this, working by males seems to be unknown in the group. Again however, it must be admitted that another explanation of the fact could be advanced: except for the faintest ambiguous suggestion in one genus there is not even any parental care by males even in the solitary nesting species, so that the evolution of worker behavior would have difficulties of initiation in this sex.

While this point must be fairly taken, nevertheless it may be that the male-haploid system is still the prime cause of the very different behavior of males.* It can be shown that it causes a selection pressure toward a sex-ratio which is markedly female-biased. This may be seen as due to the fact that in the replacement of the gene pool in each generation the females have a bigger contribution to make than the males, so that, so long as the numerical deficiency of males has not gone too far, it is

* The error mentioned in the preceding note is involved again here, as explained in the addendum (p. 87f.). In a later publication it was shown that the common female-bias of the sex ratio in Hymenoptera is likely to be due to the ability of females to store sperm.

more profitable to produce females than males. And if a chronic deficiency of males does occur it is clear that the male sex will tend to evolve adaptations for polygamous mating which must be almost completely incompatible with the evolution of male parental care. The argument concerning the sex-ratio must properly take into account the relative expensiveness of producing the two sexes. Thus if individuals all incur the same expenditure irrespective of sex, which must be the case for instance with a bee which provisions a series of cells with equal amounts of food, the ratio is the well-known ratio 1:1.618; only when a male is merely half as expensive as a female does the ratio sink to the usual 1:1. The argument does not apply, however, if there is thelytoky, polyembryony, etc., and it does not apply once a worker caste has come into existence. If worker laying takes place a more male-biased ratio should prevail.

b. *Multiple-mating and multiple-insemination in Hymenoptera.* Following these considerations of sex-ratio, however, it is not surprising to find in most solitary and even moderately social Hymenoptera that the male carries more sperm than is necessary to fill the spermatheca of a single female. Generally it seems that he carries far more than enough. Possibly only in some very highly social species is multiple-insemination *necessary* to fill the spermatheca. This is an important point in favor of our hypothesis since it predisposes to the production of the very highly intrarelated families which the male-haploid system makes possible. But to what extent, over the range of groups and species, the females actually produce such families remains a large question. The literature contains many references to multiple matings by female Hymenoptera, spread over many of the major groups of the order. How frequently such multiple mating is accompanied by a significant degree of multiple insemination, and how the phenomena are distributed with respect to incipient, advanced, or retrogressing social life are matters too wide and complex to be reviewed here. For the present it must suffice to quote the very small amount of work known to the author which bears directly on multiple insemination.

Concerning female wild bees in general, Michener *et al.* state that: "Spermathecas with only a few sperms have not been found, in spite of some search, although specimens with the spermatheca only half-full are known." But in a survey of some Australian halictines, Michener found that on the whole the number of sperms in spermathecas was small in comparison with his experience of American halictines. Without knowledge of the quantity of sperms which the male can provide or of mating behavior, one cannot be sure what this argues about multiple-insemination, but it suggests that it may be uncommon. Taken together with Michener's notable failure to find any small-ovaried worker-type bees, which according to him are a feature of most common halictines of other continents, this observation seems, therefore, against our hypothesis. But Michener notes as another general feature, the short adult life-span of the Australian bees and concludes, "There is no evidence that any female lives long enough to encounter her adult progeny," which at least offers another possible reason why worker behavior has failed to appear. Plateaux-Quénu thought queens of the quite highly social *Halictus marginatus* were probably multiply inseminated because she found some queens toward the beginning of the period of fertilization with only partially filled spermathecae. Michener and Lange present evidence that a female of the solitary (though gregarious-nesting) anthophorine bee *Paratetrapedia oligotricha* in Brazil, taken in copula, was engaged in receiving her second insemination, this apparently being the only direct evidence of such a thing in a primitive bee known to them at the time.

Multiple insemination of a high order effectively producing a progeny of multiple paternity seems to be firmly established for the honeybee. On the other hand, it would seem not to occur in the Meliponinae. It occurs in the socially very advanced fungus-growing ants, e.g., *Atta sexdens*. But in another myrmicine of a different tribe I found no evidence but of single inseminations, using Kerr's sperm counting methods.

Suppose a female is mated by n males and they are respectively responsible for proportions

$$f_1, f_2, \cdots f_s, \cdots f_n, \left(\sum_s f_s = 1\right),$$

of her female progeny. The average relationship between daughters is then

$$\tfrac{1}{2}\left(\tfrac{1}{2} + \sum_s f_s^2\right).$$

In particular, if all males contribute equally we have

$$\bar{r} = \tfrac{1}{2}\left(\tfrac{1}{2} + \frac{1}{n}\right),$$

which is the lowest average relationship for a given value of n. If two males contribute equally we have $\bar{r} = \frac{1}{2}$ as for normal full-sibs. Clearly multiple insemination will greatly weaken the tendency to evolve worker-like altruism and $n > 2$ in the model situation described above should prevent its incipience altogether. Using Taber and Wendel's estimate of $\bar{n} = 8$, which Kerr's different method roughly confirmed, we get $\bar{r} = \frac{5}{16}$, which doubtless should be raised a little to allow for inequality in the contributions of the drones. It does seem at first rather surprising that altruism toward sisters so much *less* related than full sisters can be maintained at its observed pitch of perfection. But even the limiting value of $\bar{r}$ is no lower than $\frac{1}{4}$ and we may well imagine that once established the biological advantage of the social mode of reproduction, which is evidenced by the success of the social insects in general, proves sufficient to outweigh even a two-fold higher value in personal offspring. It may also be argued that the firm establishment of highly differentiated trophogenic worker castes creates a gulf which a sexualized mutant is unlikely to cross successfully, especially when as in honeybees so much depends on the behavior of the other workers. For example, a mutant sexualized worker of the honeybee will eclose from a worker cell and will therefore be small. Even if it secretes queen-substance it is unlikely to be as attractive as a proper queen and is likely to be killed. If it escapes it cannot found a colony on its own. Thus, if the trend to multiple insemination occurs after the firm establishment of the worker caste, its threat to colonial discipline is a rather remote one.

In species of social Hymenoptera which found their colonies through single fertilized females the difficulties and dangers of seeming to be royal are less important; but the mutant will still suffer handicaps from its probable small size and lack of food reserves. In ants it will be further handicapped by its lack of wings. Nevertheless with ants there are strong indications that trends of worker sexualization have occurred in the evolution of the group. It may be remarked that the sexualized worker is likely to have a smaller spermatheca and so to restore single insemination, which will, according to our theory, restore the basis for re-evolving strong worker altruism.

An ability of females to lay unfertilized eggs which develop into females would open another possible avenue for selfish selection. Again, the menace will be greatest when multiple insemination of queens occurs, for then when a worker had inherited the causative gene from its father there would be a better chance, especially when the gene frequency was low, that it would have some normal worker sisters to help rear its offspring. In general, whether we are concerned with parthenogenetic production of males or females we need only follow Sturtevant's argument and visualize the drastic or fatal overproduction of sexual or egg-laying forms which would occur in the "son" or "daughter" colonies due to an egg-laying worker to see the potent counter-selection to which a fully penetrant causative gene will become exposed. Clearly the situation is worse for the gene when it is common than when it is rare so that an equilibrium is possible.

Female-to-female parthenogenesis by workers does occur sporadically in honeybees and shows geographical variation in its incidence. In the South African race, *Apis mellifera capensis* Esch., it seems that worker eggs always develop into females. But whether this is explicable as a selfish trait is rather doubtful. To be such the laying-workers would have to try to get their eggs cared for in queen-cells. Despite what Flanders seems to quote Onions as having stated—that in queenless hives "Uniparental workers do not construct either queen cells or drone cells," and that "a queenless colony gradually disintegrates"—Dr. Kerr informs me that these bees do eventually construct queen cells in an emergency and can thereby secure the perpetuation of their colony; but he found that they did so somewhat tardily compared to queenless colonies of the familiar honeybees. Of course for them, possessing this unusual ability, the need to initiate queen-rearing at once is not so urgent. Also in some other races diploid eggs laid in queen cells by workers in hopelessly queenless hives may sometimes be reared and so save their colonies from extinction.

Female-to-female parthenogenesis is also present in various species of ants. For example in the ant *Oecophylla longinoda*, parthenogenesis of a clonal type seems to have become a normal mode in the reproduction of the colony. Here the workers and not the mother queen produce the new generation of queens, which is suggestive at least that the situation had its origin through the selection of a selfish trait.

c. *Termites.* The special considerations which apply to the Hymenoptera do not seem to have been noticed by Williams and Williams. The discussion which they base on their analysis of the full-sib relationship would, however, be applicable to the termites where this relationship is ensured in the colony by having the queen attended by a single "king." Termites of both sexes have an equal relationship ($r = \frac{1}{2}$) to their siblings and their potential offspring. Thus the fact that both sexes "work" is just what we expect; we need only a bioeconomic argument to explain why restriction of fertility to a few members has proved most advantageous to the sibship as a whole. On this point the present theory can add little to previous discussions.

When either king or queen dies the worker castes rear a substitute or "neotene" from the eggs or young nymphs already present. The neotene mates with the surviving parent. The progeny which come from such a mating will still be related to the old workers with $r = \frac{1}{2}$. They will be related among themselves by $r = \frac{5}{8}$. They will also tend to be highly homozygous and such matings are in fact said to be somewhat infertile.

It is surprising, however, if increasing the tendencies to social cohesion by such close inbreeding can ever pay off as a long-term policy against the disadvantages of decreasing adaptive flexibility. That it may be a successful short-term policy for a species is perhaps indicated by the frequency of mention of brother-sister mating in the literature on social insects; but these statements are not always based on very firm evidence.

d. *Pleometrosis and association; population viscosity in the social insects.* However it does seem necessary to invoke at least a mild inbreeding if we are to explain some of the phenomena of the social insects—and indeed of animal sociability in general—by means of this theory. The type of inbreeding which we have in mind is that which results from a high viscosity of population or from its actual subdivision into small quasi-endogamous groups.

In some ants (e.g., *Iridomyrmex humilis*), at least one species of stingless bee (*Melipona schencki*) and apparently most species of wasps of the sub-family Polybiinae it is normal to have at least several "queens" engaged in egg-laying in each nest. This phenomenon is known as pleometrosis. Colony reproduction is by swarming with several or many fertilized females—potential queens—in each swarm. Clearly this social mode presents a problem to our theory. Continuing cycle after cycle colonies can come into existence in which some individuals are almost unrelated to one another. Such situations should be commoner the higher the number of founding queens, but less common in so far as there is any positive assortment of true sisters in the swarms. They would be very favorable to the selection of genes causing selfish behavior and this in turn would be expected to lower the efficiency of social life and to reduce the species. Yet though selfish behavior is certainly not absent—witness the large proportion of unfertilized wasps in egg-laying condition, and the common occurrence of dominance behavior—it does not seem to do the colonies much harm and the species concerned are highly successful in many cases.

For example, the genus *Polybia* includes several very abundant species in the Neotropics and has obviously undergone considerable speciation with the whole system in working order.

Wasps of the widespread genus *Polistes*, doubtfully placed in a separate sub-family from the Polybiinae, present a rather similar problem. In this case it seems that there is usually or always only a single principal egg-layer on the nest; she dominates the others and they succeed in laying only a few eggs if any. But with many of the species and races that inhabit warmer lands it is common for the initial building of the nest to be the work of two or more fertilized queen-sized wasps, a phenomenon called "association." Even at this stage the dominant wasp does least work and probably all the egg-laying, and, probably due to their more arduous and dangerous lives, the auxiliaries (as the subordinate queen-like wasps are called) tend to disappear in the course of time so that a queen assisted by her daughter workers becomes the normal situation later on. Here it is the ready acceptance of nonreproductive roles by the auxiliaries that we have difficulty in explaining. There is good reason to believe that the initial nest-founding company is *usually* composed of sisters, which brings the phenomenon closely into line with the pleometrosis of the polybiines. But it is doubtful if the wasps have any personal recognition of their sisters and if a wasp did arrive from far away it is probable that it would be accepted by the company provided it showed submission to the one or two highest ranking wasps. Dominance order does sometimes change and an accepted stranger has before it the prospect of rising in rank and ultimately subduing or driving off the queen. Thus an innocent rendering of assistance is not always easy to distinguish from an attempt at usurpation as Rau has pointed out, so that the readiness to accept "help" is really just as puzzling as the disinterested assistance which some of the auxiliaries undoubtedly do render.

The geographic distribution of the association phenomenon in *Polistes* is striking. We may state it as a general, though by no means unbroken, rule that northern species approximate to the vespine mode of colony foundation and tropical species to the polybiine to the extent above described. The single species *Polistes gallicus* illustrates the tendency well. At the northern edge of its range in Europe its females usually found nests alone. In Italy and Southern France the females found nests in companies; while in North Africa the species is said to found colonies by swarming with workers. We here suggest two hypotheses which could bring these facts into conformity with our general theory.

The first posits a general higher viscosity of the tropical populations. This will cause, through inbreeding, all coefficients of relationship to have higher actual values than we would get taking into account only connections through the past one or two generations. And it will also increase the tendency for casual neighbors to be related, which is clearly of potential importance for the association phenomenon.

Populations of *Polistes* certainly are very viscous. Generally the wasps have a strong attachment to their place of birth, and like to found nests near the parental nest. They are weak flyers. And they do show a very pronounced tendency to local variation. But whether these remarks apply any more strongly to tropical than to temperate populations I do not know. Polybiine wasps seem to be weaker flyers than vespines and also have indications of a tendency for swarms to build not far from the parent nest. Polybiines also show much geographical variation.

By its very nature the so-called temperate climate may tend to force a greater degree of vagrancy on the insects inhabiting it, both through its pronounced seasons and its seasonal irregularities. A discussion of this idea from a similar biological point of view can be found in Wynne-Edwards. As one further factor relevant at least to *Polistes* we suggest that if, as seems probable, the genus is of tropical origin the northern species will be derived from former races which themselves tended to be made up of vagrant colonist wasps which had flown north. Thus there would have been selection for wasps willing and able to found nests alone; and in general, in the course of such a spreading colonization, a species would be expected to shed some of its cooperative adaptations. But if the spread was very slow, as it may well have been, these factors would hardly apply.

The second hypothesis appeals to the lack of marked seasons in the tropics causing a lack of synchronism in the breeding activity of insects. This will tend to cause inbreeding because it scarifies the mating population. Thus a Polybiine nest may be in active production of sexual wasps when its nearest neighbors are not and its progeny may therefore be more inclined to mate among themselves. The same doubtless applies to *Polistes* in a really equable tropic environment and with *Polistes* we again have an important correlative effect that when a nest-founding wasp accepts an adventive helper the chance that she is a sister is also increased. However, with *Polistes*, multiple-queen nest founding does occur even where the wasps are constrained by the climate to follow a definite seasonal cycle. Queens may come together in the spring after hibernation to found their colonies. Rau records some interesting observations on *P. annularis* in the United States showing the variability of its nesting behavior and he mentions his general experience that the hibernated queens return to the old nest for a short time before going off to found nests. Such behavior should help to ensure that in cases of associative founding the cofoundresses are sisters. The cases of hibernating yet associating *Polistes* would seem to dismiss any hypothesis that the differences we have noted between northern and tropical wasps are due solely to factors following from the necessity for hibernation. In *Vespula*, queens do often hibernate in the parental nest and yet do not show association in nest-founding.

To the extent that they are valid, the above hypotheses would also help to extenuate previously discussed difficulties concerning the maintenance of reproductive altruism despite multiple insemination of queens. It may be remarked that although modern work rather indicates that its breeding system is far from viscous the honeybee does seem to maintain local races quite readily. With *Atta sexdens* I have noticed that males and females come to earth from their nuptial flight in local concentrations, but whether these are associated each with an established colony or represent some wider nuptial gathering is not clear.

e. *Aggressiveness.* The aggressiveness of the workers of social insects toward disturbers of their nest is one of the most conspicuous features of their altruism. The barbed sting and the function of sting autotomy are physical parallels of the traits of temperament. The correlation of these characters with sterilization does seem to hold very well throughout the social Hymenoptera. Queens are always timid and reluctant to use their stings compared to workers. In *Polistes,* workers, unless very young, are more aggressive than auxiliaries, and auxiliaries more than the reigning queen. Races of honeybees in which laying workers occur more frequently or appear more readily when the hive becomes queenless are generally milder than the races where they are less prevalent. Polybiine wasps, pleometrotic and lacking pronounced caste differences, are generally somewhat less fierce than vespines.

However, aggressiveness is also clearly a function of the size of the colony, or perhaps even more of the worker : queen ratio. This applies not only to particular colonies as they grow larger but also in a general way to variation in mature colony size between species. This effect too is not very surprising, for, to take the extreme case, we can see that it is only when its nest is overpopulated and its services in other directions superfluous that the worker can afford to throw its life away. Typically the vespines have the higher worker : queen ratio, so that from this point of view as well, it is not surprising that the polybiines are generally speaking milder wasps. It is interesting to learn that even in the limited north–south range covered by the islands of Japan, *Polistes* shows in this respect as well its previously noted tendency to bridge the two types. Yoshikawa gives an interesting comparison of northern and southern Japanese species and it is seen that northern species are both fiercer and have the larger colonies. Iwata (quoted by Yoshikawa) believes that the fierceness is a function of the colony size. Although no properly associative *Polistes* occur in Japan, Yoshikawa has found a case of temporary association in a southern species, suggesting a slight or vestigial tendency. Perhaps this factor may play a part in the difference in fierceness. Existence of auxiliaries would seem incompatible with a high degree of worker differentiation and will therefore tend to counter the development of high worker altruism. But just why it appears to be also incompatible with higher worker : queen ratios is not entirely clear.

f. *Usurpation.* Its made or half-made nest is obviously a valuable property to a queen bee, wasp, or ant. If it is ready provisioned or staffed by workers and set for the rearing of sexual brood it is even more valuable. It is therefore not surprising that usurpation has become a major evolutionary and behavioral issue with the nesting Hymenoptera.

On the one hand we have the great array of parasites. Often, especially in bees and wasps, the host and parasite species seem to be closely related, suggesting that the habit arose out of petty intra-specific usurpation. But the present theory indicates considerable difficulties for the sympatric emergence of a parasitic race. Unless the evolving complex of characters could include a strong tendency to vagrancy the usurper would in too many cases destroy the genes on which its own behavior was founded. One allopatric race invading the territory of another with at least partial reproductive barriers already present should create a more promising situation for progress in usurper-instincts. A situation like this, involving occasional parasitism, is suggested for two species of *Bombus* in Britain. Plateaux-Quénu has observed a half-provisioned nest of *Halictus marginatus* being used by a female of *H. malachurus.* Both these species are social on about the same level as *Bombus.*

On the other hand, we have the sensitivity about adventive females which is so widespread in the nesting Hymenoptera, including the parasites themselves. According to Plateaux-Quénu, conspecific usurpation is frequently attempted, albeit before the appearance of the workers, in the nest aggregations of *Halictus malachurus* and sometimes succeeds. A successful conspecific usurpation, strongly resisted, has actually been observed in *Polistes fadwigae* by Yoshikawa, and I have observed what was probably an attempt, persistent but unsuccessful, in *P. versicolor*. Something similar seems to have been seen by Kirkpatrick with *P. canadensis*. And with the same species I have found that if a dominant wasp is transferred from one nest to another a mortal fight, usually with the reigning dominant, begins immediately; whereas a young worker similarly transferred may sometimes be accepted and, perhaps because of its submissiveness, seldom receives so severe an attack. Extreme suspicion concerning wasps which approach the nest in a wavering uncertain manner sometimes prevents a genuine member of the colony from rejoining it, at least for some time, in *P. canadensis*. This is especially apt to happen with young wasps, perhaps returning from their first flight; and it may be a rather paradoxical result of such a reception that they sometimes end up working on a nearby nest not their own. Possibly it is the danger of usurpation, joint with that of parasitoids, that keeps so large a proportion of a *Polistes* colony idle on the nest when one would have thought they could be much more usefully employed out foraging.

As the very existence of association necessitates, antagonistic behavior is not so marked in the very early stages of nest-founding; then, with *Polistes versicolor*, a considerable amount of swapping of wasps may take place from week to week within a local group of initiated nests—for example all those located around the buildings of a household and usually not far from a last year's abandoned nest from which very likely all or most of the wasps are derived. The same sort of thing has been noted by Ferton for *P. gallicus* and by Rau for *P. annularis*. But even at this stage fights are sometimes seen severe enough for the combatants to fall off the nest.

In these associative *Polistes* the great variation in the degree of association —from lone nest-founding to companies of 12 or more crowded on and about a tiny nest-initial—the frequent abandonment of young nests, the quarrels, the manifest concern about adventive wasps, combine to create an impression which is very reminiscent of the breeding affairs of the South American cuckoos *Crotophaga ani* and *Guira guira* as described by Davis.

In their broad features the situations are indeed so similar as to suggest similar trends of selection must be at work in populations similarly patterned with respect to relationship. In these birds, much as in *Polistes*, we have a basic ability to rear young independently complicated by a tendency of some birds to assist altruistically (perhaps most marked in *Crotophaga*) and of others to play the cuckoo (most marked in *Guira*, which also sometimes parasitises other birds). A striking difference from *Polistes* of course is the presence of males, playing parts in close parity with those of females. And the systems also differ in that usually several birds succeed in laying in the communal nest, which is more like what is found with certain primitively social xylocopine bees than like *Polistes*. When the clutch becomes very large through this cause a large proportion of eggs may fail to hatch. Eggs are sometimes taken out and dropped. Such action by a particular bird might serve to increase the proportion of its own eggs in the clutch. For all the seeming confusion and inefficiency these birds are, like *Polistes versicolor* and *P. canadensis* in the same area, widespread and apparently successful.

g. *Pleometrosis in Halictinae*. The social halictine bees closely parallel the systems found in *Polistes*. Worker populations are of comparable size. The state of affairs found in *Augochloropsis sparsilis* and in *Lasioglossum inconspicuum* shows that these species have a class closely corresponding to the auxiliaries of warm-climate *Polistes*. But since at least some of the halictine nests are pleometrotic it seems more probable that some of their auxiliaries become layers later on rather than dying young as workers as they tend to do in *Polistes*. Probably only a minority of the species of Halictinae have any trace of a worker caste and the group also differs in the wide range of types of sociability which their tunnel-nesting encourages. For instance, quite a common situation with burrowing bees, both Halictinae and others, is for several females to be using a common entrance tunnel while each owns a separate branch tunnel further back.

Michener has recently suggested that the road to sociability and the development of a worker caste has lain in this group through a stage like this followed by a stage like that found in *Augochloropsis sparsilis*. This we are inclined to doubt since even if the nest-system users are for some reason always sisters the genetic relation of sister eggs will always be twice that of niece eggs irrespective of multiple insemination, so that on the present theory social evolution via the matrifilial colony always offers the easier route to worker altruism. Hymenopteran societies in which the queen (or queens) have auxiliaries but not, later, filial workers, seem in fact to be unknown. The classical theory concerning the evolution of the social insects has always posited a wide overlap of generations allowing mother and daughters to co-exist in the imaginal state as one of the preconditions for the evolution of this kind of sociability, and it is surely significant that it is never observed where this condition is lacking, as it might well be if genetic interest in nieces were sufficient to encourage reproductive altruism. That such altruism could arise through genetic interest in the offspring of unrelated bees sharing the same excavation, as Michener actually suggests, seems to me incredible.

h. *Tunnel-guarding by bees.* There is however another important type of social behavior to which Michener has re-drawn attention which might well arise on the basis of much lower relationships. One of the potential advantages when two or more females share a common entrance tunnel is that the entrance can be defended against parasites by a single bee, leaving the others free to forage. Instincts for guarding a narrow entrance seem to be widespread in the nest-excavating bees and also occur in Meliponinae. Michener has seen females of *Pseudogapostemon divaricatus*, a workerless but entrance sharing species, apparently taking turns at the duty and he and other observers have seen guard bees of this and other species repulse mutillids and parasite bees. The menace of intruding parasites may give such cooperation a very high advantage. But it would seem that once established the system should give an even higher advantage to the sporadic "shirker," so that it is a little difficult to see how guarding could become perfect. Perhaps it is not. One may however construct a simple imaginary system that would render it so: the bees could evolve an instinct which allowed them to leave duty at the nest entrance only on the stimulus of another recognized tenant coming in, or better, of another bee coming up from behind; this would ensure that there was always a bee on duty or at least somewhere in the nest system. By going out when supposed to be on duty, a bee would jeopardize her own brood as much as, if not more than, the broods of the others, so that selection would tend to stabilize the instinct. Interestingly Claude-Joseph and Rayment both have claimed to have observed guarding on this system, but Michener is inclined to doubt these claims because his careful observations on *P. divaricatus* in Brazil had revealed that the behavior was more irregular than might appear at first sight, bees remaining on guard for some time and allowing others to go out past them. In a highly pleometrotic nest-system, shirking might be relatively easier and safer for the isolated social deviant but the spells of guard-duty demanded would also be much shorter and therefore the selective incentive to shirking much less. Nevertheless, even if it is possible to account for the evolution of guard-instincts without a basis of relationship between the bees, it is hard to see how other socially disruptive practices, such as robbing within the nest-system, could fail to evolve unless a bee's co-tenants were also usually the carriers of some part of its inclusive fitness.

6. Equal-Status Situations

By an equal-status situation we simply mean a social situation where there is no obvious and regular difference in age, caste, or sex between the individuals concerned. Several apparently of this nature have already been mentioned, including the nest-system of independently working solitary bees discussed just previously. Now, using other examples, we indicate some other kinds of argument which it may be useful to apply to these situations.

Cooperation

In certain ants, notably *Lasius flavus* and *L. niger*, it is known that companies of several queens will cooperate in excavating the initial nest. Since these have just come to earth from a vast mating-flight they are unlikely to be close relatives. According to Waloff, the queens of *L. flavus* usually cohabit peacefully in the nest-chamber and even keep their eggs in a common pile, but about the time cocoons are first formed they tend to separate, some taking a portion of the brood (not necessarily a very fair one it seems) to a particular corner of the nest. There is evidence that the queens so separated tend to control distinct sectors in the developing nest, each having its own worker population; and whether by death of queens—by fighting or otherwise—or by migration of a "sector," most nests of *L. flavus* end up haplometrotic. In *L. niger* fighting between the queens is regular and generally only one survives in the initial nest chamber.

If we imagine a situation where, of the queens which succeed in cooperatively establishing an initial nest, only one is allowed to survive and use it, rather as happens with *L. niger* except that the survivor is chosen at random and not according to fighting prowess, we see that unrelated queens will evolve instincts to cooperate as a group of n if the chance that they succeed in establishing the chamber is more than n-times the chance that one would succeed if alone. When engaged in digging the queens are very helpless and it is not difficult to imagine that a team gets itself underground so much more quickly than an individual that this criterion is met. As to the continued amity once the chamber is made Waloff's observations on experimental multi-queened initial nests showed that for some reason the queens survive better and rear their first workers sooner when in a group than when alone; if sufficiently marked in a state of nature such an effect could explain the continued amity. With the appearance of the first workers the queen and her brood tend to become more independent and we expect behavior to change accordingly.

It will be seen that in essentials this situation has much in common with that previously described concerning the fusion of sporelings of *Gracillaria verrucosa.*

In both cases we have a strong presumption that a stage in which selection very strongly favors the united group over the lone individual gives place to conditions where the individual would be better off in the absence of its close companions. According to our theory whether these new conditions will bring on an overt struggle or fighting will depend very largely on the degree of relationship in the group in question, or rather on the degree of relationship that has held on average in the multitude of similar situations which have occurred during the evolutionary development of the behavior.

Fights

The argument to be applied to fights is merely another form of the argument applied above to cooperation. If two evenly matched unrelated animals holding one unit of reproductive potential each are in a typical situation which holds out the prospect of a fight, and if their instincts have been nicely adjusted by natural selection to suit the average outcome, then they will fight only if the expectation of reproductive potential for the winner is more than one unit. If they are sibs they will fight only if the expectation of "winner's r.p. + $\frac{1}{2}$ loser's r.p." is greater than $1\frac{1}{2}$. Thus if one inevitably dies in the fight the winner must normally gain by more than 50% or the two will prefer to co-exist. In the case of a "hymenopteran full-sistership" they will not fight to the death unless the expected gain to the winner is more than 75%. But with the honeybee, with the amount of multiple-insemination discussed previously, about a 40% increase will be enough. In this case we may put it that unless the presence of an extra young queen can increase the growth of the colony by more than 40% the reigning young queen will prefer to do without her. Thus the mutual animosity of young queens is not very surprising. The "piping" by a still imprisoned queen incidentally would seem to have the characteristics of altruism. The females of various species of Hymenoptera (in *Vespula, Halictus*, etc.) are said to fight in spring for the possession of the maternal nest in which they hibernated together. But in these cases we do not know about multiple-mating and anyway there is probably no question of a fight to the death, beaten females are usually expelled and presumably go off to discover or excavate other nest sites.

It may be noted that the larvae of "gregarious" parasitoid Hymenoptera in whose case there is normally no question of "going off" do not fight even if overcrowded. "Gregarious" refers to species where the adult normally lays several eggs per host. In "solitary" species, which lay only one egg per host, the first instar larva is adapted for fighting and always attempts to kill any other larvae in the same host; normally there is only one survivor. The gregarious larvae in a host are not necessarily "hymenopteran full-sisters" however, even apart from the question of polyandrous insemination. In cases of polyembrony they will be clonal. The same would be true if the mother reproduces by thelytoky, and, as Dr. Salt has reminded me, this certainly occurs in some ichneumon-flies; and female-to-female parthenogenesis of one kind or another is widespread throughout the parasitoidal Hymenoptera. In such cases the comparison to the batch of females competing for the nest is still less valid, although in itself the difference in social behavior between these two types of parasitoid according to relationship remains very striking. In the parasitoidal Diptera with normal sexual reproduction our theory predicts that the competition between gregarious larvae should be fiercer; whether this is observed I have been unable to ascertain, but gregarious cases are certainly much less common in Diptera and this at least is what we expect.

Parental Behavior to Minimize Sibling Competition

Of course the above argument is of potentially much wider application. It may be applied to broods of insect larvae feeding under circumstances where the exigencies of competition are not so inflexible as with parasitoids, for example to broods feeding on plants. Competition within such broods should according to our theory be fiercest in species where the female is inseminated by many males. Fierce competition will waste the energies of the brood and the ovipositing behavior of adult females should tend to evolve so as to minimize this wastage which spells a lowering of total surviving progeny. Hence over a range of species the habit of laying eggs in batches should correlate with monandrous insemination of females. The correlation should be stronger for cases where the larvae are also gregarious.

This reminds us of Fisher's suggestions concerning the evolution of distastefulness and warning-coloration and we note that he appears to have tacitly assumed that the broods he discusses would be of full sibs. Probably this is fair for his cases of Lepidoptera. But as regards the sawfly larvae which he also cites, we have all the diverse hymenopteran possibilities already mentioned, both these dependent on multiple mating and, for some species, those due to female-to-female parthenogenesis. Since we know polyandrous insemination to be a distinct possibility for the Hymenoptera it is of interest to note that D'Rozario found evidence for the gooseberry sawfly, *Nematus ribesii*, that though the males are readily polygynous the female ceases to be attractive after one mating. *N. ribesii* is a good Dzierzon-rule species and concerning the sex-ratio it is said that "females predominate." If we assume equilibrium sex-ratio and equally costly males and females the average relationship is actually a little under one-half.* In accordance with Fisher's suggested correlation the eggs are laid close together and the larvae are aposematic and fairly gregarious. However to the counter-instance admitted by Fisher—the butterfly *Anosia plexippus* which "scatters her eggs although she has solitary, inedible, conspicuous, larvae"—it will be fair to add another; the case of the moth *Panaxia dominula* which also scatters her eggs although her larvae are conspicuous

* See footnote, p. 62 and addendum, pp. 87 f.

and presumably distasteful. In the vegetation they tend to be found concentrated on the preferred food plants but are probably not truly gregarious. I have noticed in the wild that the female moth ceases to be attractive as soon as she enters copulation; thus females are probably only once mated, and the case is contrary to my suggested correlation as well as to Fisher's. But of course, though these few instances help to outline the situation, they carry little weight for or against the hypothesis.

When a brood is still under parental care the parent or parents involved will be concerned to minimize the wasteful effects of sibling competition. Their disciplinary task will be easiest if the brood is of full-sibs. In the vast majority of cases it is so, either due to monogamy or to polygamy combined with parental care by the female alone. In the unusual cases of birds where polyandry is combined with male parental care it seems that the male is always monogynous and broods a clutch given him by a single female. But in some Ratites male parental care for polymaternal broods does seem to occur; and in lekking birds there would seem to be a distinct possibility of polypaternal clutches. Doubtless many more exceptions could be found. The notable case of the polyandrous social insects has already been discussed; we merely note here that the method of rearing larvae in cells is ideal for preventing direct competition and where this method is not adopted, as in *Bombus* and the social xylocopine bees, we have added reason for expecting that the queens are effectively monandrous. Nevertheless larval competition seems to be severe in some species of *Bombus*. Although the cases where full-sibships are not the rule cannot amount to much numerically compared to the vast array of cases where they are, we do not intend to suggest that diminishment of sibling competition is the sole evolutionary *raison d'être* of permanent mating ties and bi-parental broods. The cases where the tie continues, as in many birds, from brood to brood and even sometimes until one of the mates dies are sufficient to show that other factors must be operating as well.

There are some rather puzzling cases where the parent seems deliberately to provoke competition in the brood, for example by associating more eggs with a food-supply than it could ever fully support. As just one example we have the case of *Bombus* just mentioned: in *B. agrorom* it has been found that only 30 to 40% of the eggs laid become eventual adults. Mortality is greatest in the late egg and early larval stages and cannibalism among the larvae is suggested. However, the habit of many hawks of having one more nestling than it is normally possible to rear is fairly obviously a special strategy allowing for the chance that the breeding season will turn out a good one; and explanations of a like nature may appear for the other cases eventually.

The strong tendency of plants to produce seeds of standard sizes irrespective of the size of the plant shows that how available food-reserves are apportioned between seeds is not a matter of indifference to the fitness of a plant. This is indeed just what we would expect provided the situations into which the seeds disperse are not too varied. Thus for one seed to expand selfishly at the expense of its neighbors may or may not be advantageous to the inclusive fitness of its genotype but it is almost certainly not in the interest of that of the parent plant. Wind-pollination will tend more to produce half-sibships among the seeds in an ovary than will insect-pollination. Hence according to our theory if seeds in general have genotypic control of their own growth, as they surely must to some extent, wind-pollinated plants will tend to have the more pressing difficulties in respect of uniform seed production. Hence it is rather to be expected that the situation which most lays itself open to this type of competition, the ovary with numerous closely placed ovules, will be uncommon in wind-pollinated plants. By comparison with entomophilous plants this is certainly the case, although there do seem to be a few anomalous genera, e.g., *Populus* and *Juncus*. In a great many anemophilus genera carpels or gynoecia originally with two, three, or four ovules end up, through more or less regular abortions, as one-seeded "fruits." But sometimes the seeds may nevertheless be quite closely placed, as in the pine cone, the birch "catkin," the maize cob, etc.

We note again that the selfish genes for seed growth tend to waste their powers a little not only because of the assortation due to relationship but also because of the purely chance occurrence of extreme situations where gene-replicas are largely in competition with one another. But this extra effect can only be of importance when the number of seeds in the ovary is very small. A much more important contrary factor must be the tendency of wind-borne pollen grains to arrive one by one rather than all at once as with insect pollination, so exacerbating the disciplinary problem of the wind-users. But on this point, even more than on others in the above discussion, our ideas are as yet rather unclear.

7. Anomalies

Here and there in the literature are found records of behavior where relationship is conspicuously disregarded, or harms or benefits are dispensed apparently in contravention of our principles.

However, in every case known to me it seems possible to claim either that the situation has been misinterpreted or that the observation concerns a biological error; that is, a rare occurrence in an unusual situation or something of the kind. The latter would seem to be the case for instance with the unusual cases of adult birds feeding the young of other species.

Where apparently gratuitous inter-species assistance is recorded more regularly, misinterpretation must be suspected. A nonapparent return benefit signifying a symbiosis, or some degree of positive deception signifying some sort of cuckoo-parasitism, are possibilities that should be borne in mind. For instance, it has been reported that different species of xylocopine bees of the genus *Exoneura* in Australia will sometimes pool their broods in a common nest. The finding by Michener of a seemingly very similar situation in the related genus *Allodapula*, together with signs of adaptation to parasitism by one of the species, strongly suggests that the situations Rayment has observed contain at least some mild element of parasitism.

And Michener's further finding of two species parasitic on *Exoneura*, clearly derived from the genus itself and hardly separable from it taxonomically, point the same way. These two species are not adapted for pollen-collecting and hence must be fully dependent, but at least one of the supposed parasites in *Allodapula* does collect pollen and so presumably does contribute something to the nest.

Among birds the Cuculidae are a thoroughly anomalous family as regards parental care. We have already mentioned *Guira* and *Crotophaga*. Kendiegh gives a summary of knowledge of reproductive behavior in other genera. *Geococcyx californicus* also seems to have many females laying in each nest. The two North American species of *Coccyzus* show a situation rather like that which Rayment has found in *Exoneura*. The species are reported sometimes to lay in each other's nests. But both have brooding instincts and a case has been recorded where both species incubated on the same nest.

At the level of single species we may instance the occasional exceptions to the rule that nesting Hymenoptera know their own nests and do not, even if they safely could, transfer to others. As regards the transference of workers, which seems to be not uncommon in some wasps, some cases are perhaps errors due to the powers of visual recognition not being equal to the situation. A strong basis of relationship between neighbor nests, which I believe is usual with the species of *Polistes* in which I have observed worker transference (*canadensis* and *versicolor*), would greatly reduce the selection against such errors. Then there may be situations in which transferral is really in the interests of inclusive fitness, for example if a colony is dying out, or happens to find itself with more workers than can usefully be employed on it, or if a wasp brings in food when all the larvae on its own nest are completely sated. This last explanation may perhaps apply to the cases of cross-provisioning by solitary wasps in a dense nest aggregation observed by Tsuneki, and to the cases which Deleurance has observed in *Polistes* in the wild where a worker pays visits to two nests. In birds there is a parallel of a sort in a practice of nesting guillemots and razorbills. It seems that parent birds will sometimes feed the hungriest chicks in the dense nest aggregation rather than their own.

As regards the already mentioned fostering passion shown by Emperor Penguins that have lost their chicks, some doubt as to whether the observations have been correctly interpreted would seem to remain. But taking the statements at their face value we might suggest for instance, that it has something to do with heat-conservation. Perhaps the parent penguin is so closely adapted to living with its offspring that it is, at the stage in question, at a positive disadvantage without a chick nestling in the brood-pouch. But such a situation would hardly come into being unless there were strong general relationships within the flock. We seem to need to postulate this in any case to explain some other social behavior of penguins, for example, the way Adelie Penguins parents are said to leave their young in the care of only a few adults while they go off on long fishing expeditions. On the other hand, some apparently social behavior such as the formation of the crêche in severe weather is easily interpretable as being almost entirely selfish.

ADDENDUM

Sex Ratio and Social Coefficients of Relationship under Male Haploidy

Under the male-haploid system a male expects to contribute only half as much as a female to the gene pool of distant future generations. I previously thought this would cause females to value daughters more than sons until a fairly female-biased sex ratio had been produced. But gametes from sons carry the mother's genotype in full concentration whereas gametes from daughters carry it diluted by half. This exactly offsets the difference in numbers of progeny. Thus a 1:1 population sex ratio is as basic to male haploidy as to normal reproduction. Indeed as shown in a later publication it is actually more strongly stabilized in the male-haploid case. In the amended discussion of the later paper the "golden ratio," $\frac{1}{2}(3-\sqrt{5}) : \frac{1}{2}(\sqrt{5}-1)$ or 1 : 1.618, appears only as the unstable equilibrium sex ratio of a rather unrealistic special case, and other reasons are adduced to explain why male-haploid sex ratios do so very often show excess of females.

In the context of social interactions between males and females it is now clear, first, that a distinction must be made between measurement of fitness effects in terms of expected numbers of successful gametes and measurement in terms of the total fitnesses of individuals, and second, that coefficients will not be reflexive. Neither distinction is necessary when the interacting individuals are of the same sex.

The coefficients given in Figure 2 are in general probabilities that random gametes from the individuals in question are identical by descent. They correspond to Haldane and Jayakar's coeffient ϕ_{AB} (*J. Genet.*, 58, 81–107 [1962]). My own rules for the calculation of this coefficient are certainly wrong. The rules of Haldane and Jayakar also fail in certain cases, so I give below a new set based on their rules:

1. List all paths by which A and B are connected through latest common ancestors.
2. Reject any path having two males in succession.
3. Count the steps in each remaining path, treating any double steps "female-male-female" (whether apical or not) as single steps, and missing out steps to terminal males if any.
4. If n_i is the number of steps counted in the i-th path from A to B, and ϕ_i is the coefficient of inbreeding of the latest common ancestor (zero if male) on this path, then the coefficient of relationship is given by

$$\phi_{AB} = \sum_i 2^{-(n_i+1)} (1+\phi_i).$$

The coefficients apply directly on a basis of gamete-for-gamete measurement for actions of males toward females, and on a fitness-for-fitness basis they apply for actions of females toward males. In the remaining contexts, gamete-for-gamete female toward male and fitness-for-fitness male toward female, the values of the coefficients must be doubled.

The kinds of sacrifices which have to be considered as possible among social insects, such as costs of rearing larvae and the sacrifice of one life in the defense of others, are most reasonably viewed in terms of fitness-for-fitness measurement of effects. Thus the possibility of alteration of the argument brought in by the present correction mainly concerns actions of males toward females. The unit coefficients appearing for the son to mother and the son to daughter relationships are the most striking changes. At first sight they may even seem paradoxical. In the realm of theory however it is clearly true that if a son were to sacrifice his life to rejuvenate completely a dying mother he could expect the genes that he loses to be replaced. In practice he has no possibility of doing this. On the other hand there are plenty of ways in which the male can attempt to increase his production of daughters. The theoretical implication of the unit coefficient with daughters is also plain, and if the male does not exactly sacrifice his life for a daughter he does normally use it up in efforts which, in a stable population, gain him the expectation of one daughter. This unit coefficient, it must be noted, does not imply any special likelihood for the evolution of parental care. On the contrary, that the males' efforts are always toward polygamy rather than parental care is as expected. In parental care haploid males would be liable to waste some effort completely unless they were able to discriminate diploid from haploid larvae among the progeny of their mates. With normal males, at least in respect of autosomal genotype, discrimination is unnecessary.

II

Adaptive Regulation of Population Density

The appearance of Wynne-Edwards' *Animal Dispersion in Relation to Social Behaviour*, and the reactions that it produced, made group selection recognizable in this decade as a major scientific controversy. In the book the author defends his thesis in 653 tightly filled pages. The following brief article therefore can scarcely be considered sufficient documentation of Wynne-Edwards' point of view. Hopefully it will serve to illustrate the kind of reasoning used and the kind of evidence that must be interpreted in evaluating the theory.

The rebuttals by Smith and Perrins are illustrative of the criticisms offered. During the past several years many field studies of animal populations have been influenced by Wynne-Edwards' book. Some seem to support, others to refute, the concept of group selection in the evolution of social behavior.

4: *Intergroup Selection in the Evolution of Social Systems*

V. C. WYNNE-EDWARDS

In a recent book I advanced a general proposition which may be summarized in the following way: (1) Animals, especially in the higher phyla, are variously adapted to control their own population densities. (2) The mechanisms involved work homeostatically, adjusting the population density in relation to fluctuating levels of resources; where the limiting resource is food, as it most frequently is, the homeostatic system prevents the population from increasing to densities that would cause overexploitation and the depletion of future yields. (3) The mechanisms depend in part on the substitution of conventional prizes—namely, the possession of territories, homes, living space and similar real property, or of social status as the proximate objects of competition among the members of the group concerned, in place of the actual food itself. (4) Any group of individuals engaged together in such conventional competition automatically constitutes a society,

From *Nature*, 200 (1963), 623–626.

all social behavior having sprung originally from this source.

In developing the theme it soon became apparent that the greatest benefits of sociality arise from its capacity to override the advantage of the individual members in the interests of the survival of the group as a whole. The kind of adaptations which make this possible, as explained more fully here, belong to and characterize social groups as entities, rather than their members individually. This in turn seems to entail that natural selection has occurred between social groups as evolutionary units in their own right, favoring the more efficient variants among social systems wherever they have appeared, and furthering their progressive development and adaptation.

The general concept of intergroup selection is not new. It has been widely accepted in the field of evolutionary genetics, largely as a result of the classical analysis of Sewall Wright. He has expressed the view that "selection between the genetic systems of local populations of a species . . . has been perhaps the greatest creative factor of all in making possible selection of genetic systems as wholes in place of mere selection according to the net effects of alleles." Intergroup selection has been invoked also to explain the special case of colonial evolution in the social insects.

In the context of the social group a difficulty appears, with selection acting simultaneously at the two levels of the group and the individual. It is that the homeostatic control of population density frequently demands sacrifices of the individual; and while population control is essential to the long-term survival of the group, the sacrifices impair fertility and survivorship in the individual. One may legitimately ask how two kinds of selection can act simultaneously when on fundamental issues they are working at cross purposes. At first sight there seems to be no easy way of reconciling this clash of interests; and to some people consequently the whole idea of intergroup selection is unacceptable.

Before attempting to resolve the problem it is necessary to fill in some of the background and give it clearer definition. The survival of a local group or population naturally depends among other things on the continuing annual yield of its food resources. Typically, where the tissues of animals or plants are consumed as food, persistent excessive pressure of exploitation can rather quickly overtax the resource and reduce its productivity, with the result that yields are diminished in subsequent years. This effect can be seen in the overfishing of commercial fisheries which is occurring now in different parts of the world, and in the overgrazing of pastoral land, which in dry climates can eventually turn good grassland into desert. Damage to the crop precedes the onset and spread of starvation in the exploiting population, and may be further aggravated by it. The general net effect of overpopulation is thus to diminish the carrying-capacity of the habitat.

In natural environments undisturbed by man this kind of degradation is rare and exceptional: the normal evolutionary trend is in the other direction, toward building up and sustaining the productivity of the habitat at the highest attainable level. Predatory animals do not in these conditions chronically depress their stocks of prey, nor do herbivores impair the regeneration of their food-plants. Many animals, especially among the larger vertebrates (including man), have themselves virtually no predators or parasites automatically capable of disposing of a population surplus if it arises.

It is the absence or undependability of external destructive agencies that makes it valuable, if not in some cases mandatory, that animals should be adapted to regulate their own numbers. By so doing, the population density can be balanced around the optimum level, at which the highest sustainable use is made of food resources.

Only an extraordinary circumstance could have concealed this elementary conclusion and prevented our taking it immediately for granted. Some eight or more thousand years ago, as neolithic man began to achieve new and greatly enhanced levels of production from the land through the agricultural revolution, the homeostatic conventions of his hunting ancestors, developed there as in other primates to keep population density in balance with carrying capacity, were slowly and imperceptibly allowed to decay. We can tell this from the centrally important place always occupied by fertility-limiting and functionally similar conventions in the numerous stone-age cultures which persisted into modern times. Since these conventions disappeared nothing has been acquired in their place: growing skills in resource development have, except momentarily, always outstripped the demands of a progressively increasing population; there has consequently been no effective natural selection against a freely expanding economy. So far as the regulation of numbers is concerned, the human race provides a spectacular exception to the general rule.

A secondary factor, tending to obscure the almost universal powers possessed by animals for controlling their numbers, is our everyday familiarity with insect and other pests which appear to undergo uncontrolled and sometimes violent fluctuations in abundance. In fact, human land-use practices seldom leave natural processes alone for any length of time: vegetation is gathered, the ground is tilled, treated, or irrigated, single-species crops are planted and rotated, predators and competitors are destroyed, and the animals' regulating mechanisms are thereby, understandably, often defeated. In the comparably drastic environmental fluctuations of the polar and desert regions, similar population fluctuations occur without human intervention.

The methods by which natural animal populations curb their own increase and promote the efficient exploitation of food resources include the control of recruitment and, when necessary, the expulsion and elimination of unwanted surpluses. The individual member has to be governed by the homeostatic system even when, as commonly happens, this means his exclusion from food in the midst of apparent plenty, or detention from reproduction when others are breeding. The recruitment rate must be determined by the contemporary relation between population density and resources; under average conditions, therefore, only part of the potential fecundity of the group needs to be realized in a given year or generation.

This is a conclusion amply supported by the results of experiments on fecundity versus density in laboratory populations, in a wide variety of animals including Crustacea, insects, fish, and mammals; and also by field data from natural populations. But it conflicts with the assumption, still rather widely made, that under natural selection there can be no alternative to promoting the fecundity of the individual, provided this results in his leaving a larger contribution of surviving progeny to posterity. This assumption is the chief obstacle to accepting the principle of intergroup selection.

One of the most important premises of intergroup selection is that animal populations are typically self-perpetuating, tending to be strongly localized and persistent on the same ground. This is illustrated by the widespread use of traditional breeding sites by birds, fishes, and animals of many other kinds; and by the subsequent return of the great majority of experimentally marked young to breed in their native neighborhood. It is true of nonmigratory species, for example the more primitive communities of man; and all the long-distance two-way migrants that have so far been experimentally tagged, whether they are birds, bats, seals, or salmon, have developed parallel and equally remarkable navigating powers that enable them to return precisely to the same point, and consequently preserve the integrity of their particular local stock. Isolation is normally not quite complete, however. Provision is made for an element of pioneering, and infiltration into other areas; but the gene-flow that results is not commonly fast enough to prevent the population from accumulating heritable characteristics of its own. Partly genetic and partly traditional, these differentiate it from other similar groups.

Local groups are the smallest racial units capable of continuous existence for long enough to undergo evolutionary differentiation. In the course of generations some die out; others survive, and have the opportunity to spread into new or vacated ground as it becomes available, themselves subdividing as they grow. In so far as the successful ones take over the habitat left vacant by the unsuccessful, the groups are in a relation of passive competition. Their survival or extinction is partly a matter of chance, arising from various forms of *force majeure*, including secular changes in the environment; for the rest it is determined, in general terms, by heritable qualities of fitness.

Gene-frequencies within the group may alter as time passes, through gene-flow, drift (the Sewall Wright effect), and selection at the individual level. Through the latter, adaptations to local conditions may accumulate. Population fitness, however, depends on something over and above the heritable basis that determines the success as individuals of a continuing stream of independent members. It becomes particularly clear in relation to population homeostasis that social groups have highly important adaptive characteristics in their own right.

When the balance of a self-regulating population is disturbed, for example by heavy accidental mortality, or by a change in food-resource yields, a restorative reaction is set in motion. If the density has dropped below the optimum, the recruitment rate may be increased in a variety of ways, most simply by drawing on the reserve of potential fecundity referred to earlier, and so raising the reproductive output. Immigrants appearing from surrounding areas can be allowed to remain as recruits also. If the density has risen too high, aggression between individuals may build up to the point of expelling the surplus as emigrants; the reproductive rate may drop; and mortality due to social stress (and in some species cannibalism) may rise. These are typical examples of density-dependent homeostatic responses.

Seven years of investigation of the population ecology of the red grouse (*Lagopus scoticus*) near Aberdeen, by the Nature Conservancy Unit of Grouse and Moorland Ecology, have revealed many of these processes at work. Their operation in this case depends to a great extent on the fact that individual members of a grouse population living together on a moor, even of the same sex, are not equal in social status. Some of the cock birds are sufficiently dominant to establish themselves as territory owners, parceling out the ground among them and holding sway over it, with a varying intensity of possessiveness, almost the whole year round. During February-June their mates enjoy the same established status.

In the early dawn of August and September mornings, after a short and almost complete recess, the shape of a new territorial pattern begins to be hammered out. In most years this quickly identifies a large surplus of males, old and young, which are not successful in securing any part of the ground, and consequently assume a socially inferior status. They are grouped with the hens at this stage as unestablished birds; and day by day their security is so disturbed by the dawn aggressive stress that almost at once some begin to get forced out, never to return. By about 8 A.M. each day the passion subsides in all but the most refractory territorial cocks, after which the moor reverts to communal ground on which the whole population can feed freely for the rest of the day. As autumn wears on and turns to winter, the daily period of aggression becomes fiercer and lasts longer; birds with no property rights have to feed at least part of the time on territorial ground defended by owners that may at any moment chase them off. More and more are driven out altogether; and since they can rarely find a safe nook to occupy elsewhere in the neighborhood, they become outcasts, and are easily picked up by hawks and foxes, or succumb to malnutrition. Females are included among those expelled; but about February the remaining ones begin to establish marital attachments; and at the same time, quite suddenly, territories are vigorously defended all day. Of the unestablished birds still present in late winter, some achieve promotion by filling the gaps caused by casualties in the establishment. Some may persist occasionally until spring; but unless a cock holds a territory exceeding a minimum threshold capacity, or a hen becomes accepted by a territorially qualified mate, breeding is inhibited.

Territories are not all of uniform size, and on average the largest are held by the most dominating cocks. More important still, the average territory size changes from year to year, thus varying the basic population density, apparently in direct response to changes in productivity of the staple food-plant, heather (*Calluna vulgaris*). As yet this productivity has been estimated only by subjective methods; but significant mutual correlations have been established between annual average values for body-weight of adults, adult survival, clutch-size, hatching success, survival of young, and, finally, breeding density the following year. As would be expected with changing densities, the size of the autumn surplus, measured by the proportions of unestablished to established birds, also varies from year to year.

There are increasing grounds for concluding that this is quite a typical organization, so far as birds are concerned, and that social stratification into established and unestablished members, particularly in the breeding season, is common to many other species, and other classes of animals. In different circumstances the social hierarchy may take the form of a more or less linear series or peck-order. Hierarchies commonly play a leading part in regulating animal populations; not only can they be made to cut off any required proportion of the population from breeding, but also they have exactly the same effect in respect of food when it is in short supply. According to circumstances, the surplus tail of the hierarchy may either be disposed of or retained as a nonparticipating reserve if resources permit.

It is not necessary here to explore in detail the elaborate patterns of behavior by which the social hierarchy takes effect. The processes are infinitely varied and complex, though the results are simple and functionally always the same. The hierarchy is essentially an overflow mechanism, continuously variable in terms of population pressure on one hand, and habitat capacity on the other. In operation it is purely conventional, prescribing a code of behavior. When a more dominant individual exerts sufficient aggressive pressure, usually expressed as threat although frequently in some more subtle and sophisticated form, his subordinates yield, characteristically without physical resistance or even demur. It may cost them their sole chance of reproduction to do so, if not their lives. The survival of the group depends on their compliance.

This has been taken as an example to illustrate one type of adaptation possessed by the group, transcending the individuality of its members. It subordinates the advantage of particular members to the advantage of the group; its survival value to the latter is clearly very great. The hierarchy as a system of behavior has innumerable variants in different species and different phyla, analogous to those of a somatic unit like the nervous or vascular system. Like them, it must have been subject to adaptive evolutionary change through natural selection; yet it is essentially an "organ" of a social group, and has no existence if the members are segregated.

A simple analogy may possibly help to bring out the significance of this point. A football team is made up of players individually selected for such qualities as skill, quickness, and stamina, material to their success as members of the team. The survival of the team to win the championship, however, is determined by entirely distinct criteria, namely, the tactics and ability it displays in competition with other teams, under a particular code of conventions laid down for the game. There is no difficulty in distinguishing two levels of selection here, although the analogy is otherwise very imperfect.

The hierarchy is not the only characteristic of this kind. There are genetic mechanisms, such as those that govern the optimum balance between recombination and linkage, in which the benefit is equally clearly with the group rather than the individual. Without leaving the sphere of population regulation, however, we can find a wide range of vital parameters, the optima of which must similarly be determined by intergroup selection. Among those discussed at length in the book already cited are (1) the potential life-span of individuals and, coupled with it, the generation turnover rate; (2) the relative proportions of life spent in juvenile or nonsexual condition (including diapause) and in reproduction; (3) monotely (breeding only for a single season) versus polytely; (4) the basal fecundity-level, including, in any one season, the question of one brood versus more than one.

These and similar parameters, differing from one species or

class to another, are interconnected. Their combined effects are being summed over the whole population at any one time and over many generations in any given area. It is the scale of the operation in time and space that precludes an immediate experimental test of group selection. An inference that may justifiably be drawn, however, is that maladjustment sufficient to interfere persistently with the homeostatic mechanism must either cause a progressive decline in the population or, alternatively, a chronic overexploitation and depletion of food resources; in the end either will depopulate the locality.

There still remains the central question as to how an immediate advantage to the individual can be suppressed or overridden when it conflicts with the interests of the group. What would be the effect of selection, for example, on individuals the abnormal and socially undesirable fertility of which enabled them and their hereditary successors to contribute an ever increasing share to future generations?

Initially, groups containing individuals like this that reproduced too fast, so that the over-all recruitment rate persistently tended to exceed the death-rate, must have repeatedly exterminated themselves in the manner just indicated, by overtaxing and progressively destroying their food resources. The earliest adaptations capable of protecting the group against such recurrent disasters must necessarily have been very ancient; they may even have been acquired only once in the whole of animal phylogeny, and in this respect be comparable with such basic morphological elements as the mesoderm, or, perhaps, the coelom. Once acquired, the protective adaptations could be endlessly varied and elaborated. It is inherently difficult to reconstruct the origin of systems of this kind; but genetic mechanisms exist which could give individual breeding success a low heritability, or, in other words, make it resistant to selection. This could be relatively simply achieved, for example, if the greatest success normally attached to heterozygotes for the alleles concerned, creating the stable situation characteristic of genetic homeostasis.

A more complex system can be discerned, as it has developed

in many of the higher vertebrates where the breeding success of individuals is very closely connected with social status. This connection must necessarily divert an enormous additional force of selection into promoting social dominance, and penalizing the less fortunate subordinates in the population that are prevented from breeding or feeding, or get squeezed out of the habitat. Yet it is self-evident that the conventional codes under which social competition is conducted are in practice not jeopardized from this cause: selection pressure, however great, does not succeed in promoting a general recourse to deadly combat or treachery between rivals, nor does it, in the course of generations, extinguish the patient compliance of subordinates with their lot.

The reason appears to be that social status depends on a summation of diverse traits, including virtually all the hereditary and environmental factors that predicate health, vigor, and survivorship in the individual. While this is favorable to the maintenance of a high-grade breeding stock, and can result in the enhancement through selection of the weapons and conventional adornments by which social dominance is secured, dominance itself is again characterized by a low heritability, as experiments have shown. In many birds and mammals, moreover, individual status, quite apart from its genetic basis, advances progressively with the individual's age. Not only are the factors that determine social and breeding success numerous and involved, therefore, but the ingredients can vary from one successful individual to the next. A substantial part of the gene pool of the population is likely to be involved and selection for social dominance or fertility at the individual level correspondingly dissipated and ineffective, except in eliminating the substandard fringe.

Such methods as these which protect group adaptations, including both population parameters and social structures, from short-term changes, seem capable of preventing the rise of any hereditary tendency toward antisocial self-interest among the members of a social group. Compliance with the social code can be made obligatory and automatic, and it probably is so in almost all animals that possess social homeostatic systems at all. In at least some of the mammals, on the contrary, the individual has been released from this rigid compulsion, probably because a certain amount of intelligent individual enterprise has proved advantageous to the group. In man, as we know, compliance with the social code is by no means automatic, and is reinforced by conscience and the law, both of them relatively flexible adaptations.

There appears therefore to be no great difficulty in resolving the initial problem as to how intergroup selection can override the concurrent process of selection for individual advantage. Relatively simple genetic mechanisms can be evolved whereby the door is shut to one form of selection and open to the other, securing without conflict the maximum advantage from each; and since neighboring populations differ not only in genetic system but in population parameters (for example, mean fecundity) and in social practices (for example, local differences in migratory behavior in birds, or in tribal conventions among primitive men), there is no lack of variation on which intergroup selection can work.

Group Selection and Kin Selection: A Rejoinder

J. MAYNARD SMITH

Wynne-Edwards has argued persuasively for the importance of behavior in regulating the density of animal populations, and has suggested that since such behavior favors the survival of the group and not of the individual it must have evolved by a process of group selection. It is the purpose of this communication to consider how far this is likely to be true.

The strongest arguments for believing that conventional behavior is the immediate cause regulating population density concern cases of territorial behavior, particularly in birds. But it does not follow that such behavior has evolved by group selection, because territorial behavior capable of adjusting the population density to the available food supply could evolve by selection acting at the level of the individual rather than of the group. The appropriate degree of aggression would evolve if: (1) individuals which are too aggressive raise fewer offspring, either because they suffer physical damage or because they waste in display time and energy which should be spent in raising

From *Nature*, 201 (1964), 1145–1147.

their young; (2) individuals which are too timid either fail to establish a territory or establish one too small to contain an adequate food supply for the young. Further, the degree of "choosiness"—that is, the readiness to fight for a territory in one kind of area rather than put up with one in a less favorable area—will evolve by individual selection in such a way as to lead to an efficient distribution in space. This will happen because if, on one hand, individuals are too "choosy," territories in the favored areas will become too small in relation to the food supply, so that less choosy individuals breeding in the less favored but more sparsely inhabited areas will leave more offspring, whereas if individuals are too little choosy, selection will act in the reverse direction.

Thus there is no need to invoke group selection to explain the evolution of individual breeding territories, or the adjustment of territory size to food supply or to variations in the habitat. But there are other characteristics of animals which are more difficult to explain by individual selection; sex is an obvious and important example, but difficulties also arise in explaining the evolution of "altruistic" characters, such as alarm notes or injury-feigning in birds.

It is possible to distinguish two rather different processes, both of which could cause the evolution of characteristics which favor the survival, not of the individual, but of other members of the species. These processes I will call kin selection and group selection, respectively. Kin selection has been discussed by Haldane and by Hamilton.

By kin selection I mean the evolution of characteristics which favor the survival of close relatives of the affected individual, by processes which do not require any discontinuities in population breeding structure. In this sense, the evolution of placentae and of parental care (including "self-sacrificing" behavior such as injury-feigning) are due to kin selection, the favored relatives being the children of the affected individual. But kin selection can also be effective by favoring the siblings of the affected individuals (for example, sterility in social insects, inviability of cotton hybrids due to the "corky" syndrome) and presumably by favoring more distant relatives. There will be more opportunities for kin selection to be effective if relatives live together in family groups, particularly if the population is divided into partially isolated groups. But such partial isolation is not essential. In kin selection, improbable events are involved only to the extent that they are in all evolutionary change—in the origin of genetic differences by mutation.

If groups of relatives stay together, wholly or partially isolated from other members of the species, then the process of group selection can occur. If all members of a group acquire some characteristic which, although individually disadvantageous, increases the fitness of the group, then that group is more likely to split into two, and in this way bring about an increase in the proportion of individuals in the whole population with the characteristic in question. The unit on which selection is operating is the group and not the individual. The only difficulty is to explain how it comes about that all members of a group come to have the characteristic in the first place. If genetically determined, it presumably arose in a single individual. It cannot be pictured as spreading to all members of a group by natural selection, because if it could do that, it could equally well spread in a large population—either by individual selection or kin selection—and there is no need to invoke a special mechanism of group selection to explain it. Hence the only way in which such a characteristic could spread to all members of a group would be by genetic drift. (There is also the possibility that it might spread through a group by cultural transmission, but this is unlikely to be important in animals other than man.) If this were to happen at all often, then the groups must be small (or else commonly re-established by single fertilized females or single pairs), the disadvantage of the characteristic to the individual slight, and the gene flow between groups small, because every time a group possessing the socially desirable characteristic is "infected" by a gene for antisocial behavior, that gene is likely to spread through the group. These conditions are severe, although they may sometimes be satisfied.

The distinction between kin selection and group selection as here defined is that for kin selection the division of the population into partially isolated breeding groups is a favorable but not an essential condition, whereas it is an essential condition for group selection, which depends on the spread of a characteristic to all members of a group by genetic drift.

Wynne-Edwards points out that birds may return after migration to the precise spot where they were raised, and argues that this would favor the operation of group selection. This is not so. What is required for group selection is that the species should be divided into a large number of local populations, within which there is free interbreeding, but between which there is little gene flow. The mere fact that many birds breed near where they were born does not bring about this situation; it would favor the operation of kin selection, but it is difficult to see how kin selection could bring about the evolution of many of the types of population-regulating behavior which Wynne-Edwards believes he has discovered.

Wynne-Edwards also argues that the behavioral mechanisms he hypothesizes would be proof against the occurrence by mutation and subsequent spread of antisocial behavior patterns because of genetic homeostasis. This is a piece of special pleading: it also shows a misunderstanding of the situations in which homeostasis of this kind is to be expected. Both genetical theory and the experimental evidence suggest that if natural selection has been pushing a character in a given direction for a long time, it will be difficult for selection to produce further change in the same direction, but comparatively easy to produce a change in the reverse direction. Thus it would only be plausible to suggest that there are genetic reasons why antisocial behavior should not increase if it were also suggested that selection had already produced an extreme degree of antisocial behavior, and this is precisely what Wynne-Edwards denies. In fact, "antisocial" mutations will occur, and any plausible model of group selection must explain why they do not spread.

There is one special form of group selection which is worth considering in more detail, because it can, perhaps, explain the evolution of "self-sterilizing" behavior; that is, behavior which leads an individual not to breed in circumstances in which other members of the species are breeding successfully. (This is quite different from behavior which leads individuals not to breed when other members of the species are attempting unsuccessfully to breed, or to produce fewer offspring when conditions are such that they would be unable to raise a larger number; such behavior, although of great interest, presents no special difficulty to a selectionist.) The difficulty is that if the difference between breeders and nonbreeders is genetically determined, then it is the breeders whose genotype is perpetuated.

A possible explanation is that what is inherited is the level of responsiveness to the presence of other breeding individuals. Thus suppose that there are aggressive *A* individuals which continue to breed or to attempt to breed at high densities, and timid *a* individuals, which are discouraged from breeding when the density of breeding individuals reaches a certain level, the difference between *A* and *a* being genetically determined. In a mixed group of *A* and *a* individuals, if the density is high, only *A* will breed, and *a* will be lost from the group. In a group of *A* individuals at high density all will attempt to breed, with the consequence that the food supply may be exhausted and the group produce few progeny. In a group consisting entirely of *a* individuals, at high densities some will breed and some will not, the difference between breeding and nonbreeding individuals being due to age, to previous environmental history, or even to chance. Consequently an *a* group is less likely to outstrip its food supply, and so will leave more progeny. The difference between *A* and *a* groups at high densities is an example of the difference between a scramble (*A*) and a contest (*a*).

Given such a behavioral difference, the following conditions seem necessary if *a* is to increase under natural selection:

(1) Groups must for a time be reproductively isolated, because *a* is eliminated from mixed groups.

(2) Groups must be started by one or a few founders, since otherwise groups consisting entirely of *a* individuals would never come into existence.

(3) When a group of A individuals outstrips its food supply, it must not immediately encroach on the food supply of neighboring a groups, for if it did so, the advantage of a groups would disappear. This is a difficult condition to meet, and appears to rule out this mechanism in cases in which the population is divided into herds, flocks, troops, or colonies, each group having a joint feeding territory which borders that of neighboring groups. The condition is most likely to be met when the food supply is discontinuous in space, each patch of food supporting its own group.

A greatly oversimplified model of this type of selection will now be given. To fix ideas, suppose that there exists a species of mouse which lives entirely in haystacks. A single haystack is colonized by a single fertilized female, whose offspring form a colony which lives in the haystack until next year, when new haystacks are available for colonization. At this time, mice migrate, and may mate with members of other colonies before establishing a new colony. The population consists of aggressive A and timid a individuals, timidity being due to a single Mendelian recessive; a/a are timid, and A/a and A/A aggressive.

Only when a colony is started by an a/a female fertilized by an a/a male will it consist finally of a individuals; all other colonies will lose the a gene by selection, and come to consist entirely of A individuals. Thus at the time when colonies are about to break up, there are only two kinds of colony, A and a. It is assumed that an a colony contributes $1 + K$ times as many mice to the migrating populations as does an A colony, and has a proportionately greater chance of having a daughter colony.

In one summer, let the frequency of a colonies be P_0. Then, in the migrating population, the proportion of a/a individuals is:

$$\frac{P_0(1 + K)}{P_0(1 + K) + 1 - P_0} = \frac{P_0(1 + K)}{1 + KP_0} = p \text{ say}.$$

It is assumed that a proportion r of all migrating female mice mate with males from their own colony, the remaining $(1 - r)$

mating at random. Hence the frequency of $a/a \times a/a$ mating as a fraction of all matings is

$$rp + (1 - r)p^2 = P_1$$

where P_1 is the frequency of a colonies in the next summer.

Hence the condition for the evolutionary spread of "timid" behavior—that is, of the a gene—is:

$$rp + (1 - r)\, p^2 > P_0 \text{ where } p = \frac{P_0(1+K)}{1+KP_0}$$

This reduces to

$$r(1 + K) - (1 - P_0K^2) > 0$$

Thus when P_0 is large ($P_0 \simeq 1$),

$$r + K > 1$$

and when P_0 is small ($P_0 \simeq 0$),

$$r(1 + K) > 1$$

Thus, if there is little or no interbreeding between colonies even at migration ($r \simeq 1$), timid behavior will evolve provided it is an advantage to the group; this corresponds to the case in which the population is divided into more or less permanently isolated groups, which are periodically reduced to very small numbers, and which may either become extinct or split to give rise to two groups. However, the conclusion that timid or altruistic behavior can readily evolve if there is no interbreeding between groups means little, since it is unlikely that species are often divided into a large number of small and completely isolated groups.

If there is fairly free interbreeding between colonies at regular intervals (that is, if r is small), selection could maintain the gene for timidity once it had become the common allele in the population. For example, if there were random mating, $r = 0$, between members of different colonies at the time of migration, then selection could maintain a as the common allele if a colonies had a two-fold advantage. But, with random mating, selection could not cause a to increase if it were initially rare: if $r = 0$, the condition $r(1 + K) > 1$ cannot be satisfied.

With an intermediate amount of gene flow between colonies, selection could both establish and maintain timid or altruistic behavior, provided that colonies with altruistic behavior have a large selective advantage, and that colonies are founded by very few individuals.

The model is too artificial to be worth pursuing further. It is concluded that if the admittedly severe conditions listed here are satisfied, then it is possible that behavior patterns should evolve leading individuals not to reproduce at times and in circumstances in which other members of the species are reproducing successfully. Whether this is regarded as an argument for or against the evolution of altruistic behavior by group selection will depend on a judgment of how often the necessary conditions are likely to be satisfied.

A Reply to Maynard Smith's Rejoinder

V. C. WYNNE-EDWARDS

Dr. Maynard Smith's communication raises a good many more questions than I can attempt to answer here. The major obstacle to constructive discussion between us really arises from the understandable (though regrettable) differences in outlook and experience between a laboratory geneticist and a field ecologist. To me his picture of territorial systems and other aspects of conventional behavior appears scarcely true or comprehensive enough to provide a basis for valid deduction; my own grasp of the genetical theory of natural selection, on the other hand, no doubt looks still more halting and inept to him. We ought to enlarge the area of common ground, but that is too big a task to discharge effectively here.

It is not permissible to isolate "territory" in Dr. Smith's sense from the other overlapping forms of real property won and defended by animals, such as nest-sites of colonial birds, basking or resting places of, say, seals or crocodiles, roosting

From *Nature*, 201 (1964), 1147.

perches of starlings or domestic fowls, display arenas of manakins or bower-birds, burrows of foxes or beach crabs, and so on. Many of these have no direct connection with food or with rearing families; but all are indissociably bound up with the status of their possessor in the social system to which he belongs and the rights which this status confers. What we have to explain is how social systems can evolve and their conventional machinery be perfected. What appears to be inevitably required is a process of selection discriminating between one social system and another.

Social systems are collective entities, in the higher animals frequently involving an element of tradition as well as genetic transmission as they pass down from generation to generation. They entail codes of behavior with which the individual members instinctively comply, even when compliance demands the resignation of rights to vital resources or to reproduction. The hereditary compulsion to comply, for example, in lemmings doomed to emigrate or sticklebacks inhibited from maturing by the inescapable domination of an α male, is the real keystone of social adaptation. Individuals submitting to total deprivation are eliminated altogether, most often before they have produced any offspring; yet the tendency to comply is renewed in every subsequent generation and is not bred out. One is bound to conclude that it is very securely buffered from "ordinary" selection acting against submissive individuals and at the same time promoting their dominant sibs; and from the effects of simple Mendelian situations of the *A*/*a* type in Dr. Smith's model. I stand corrected if it is technically wrong to think of this as genetic homeostasis; the apparent result is the same. The situation I describe here is real and not, I think, controversial; it is the explanation which presents difficulties.

Most ecologists would agree that the prerequisite of group selection that calls for a subdivided population structure is commonly and indeed normally found in animals. Dr. Smith says that the *Ortstreue* or return of migrant birds to their native locality would not bring it about; perhaps it is easier to see then in the case of the salmon or trout spawning in its natal tributary stream, where it more obviously becomes a member of a partially isolated breeding group.

The model of the mice in the haystacks is not, perhaps, a sufficiently close approximation to any natural situation to help us far toward a solution. A realistic counterpart might be, for example, the woodlice (*Porcellio scaber*) that fed on the green alga *Protococcus* living on tree-trunks, studied by Brereton; marked woodlice confined their feeding to their own particular tree, and the population appeared to be subdivided thus into breeding units. Had any of the latter increased too freely they could have exterminated their stock of this particular food plant, which does not regenerate easily. Supposing in Dr. Smith's model that all the A colonies grow so fast that they finish the food and die of starvation before "migration time" arrives; then $K = \infty$, $p = 1$, $r = 1$, and $P_1 = 1$, and group selection wins the trick!

Survival of Young Swifts in Relation to Brood-Size

CHRISTOPHER PERRINS

In a recent article in *Nature*, Wynne-Edwards said "the assumption, still rather widely made, [is] that under natural selection there can be no alternative to promoting the fecundity of the individual, providing this results in his leaving a larger contribution of progeny to posterity. This assumption is the chief obstacle to accepting the principle of intergroup selection."

Wynne-Edwards implies that this assumption is wrong, but omits to note that there are data which show that at least some species are producing as many surviving young as possible, for example, the starling, *Sturnus vulgaris*, the great tit, *Parus major*, and the laysan albatross, *Diomedea immutabilis*. The swift (*Apus apus*) is a convenient species for a study of this kind since, like the albatross but unlike the passerine species studied, considerable mortality, due to starvation, occurs in the nest.

From *Nature*, 201 (1964), 1147–1148.

In England the swift normally lays a clutch of two or three eggs, clutches of four being very rare indeed (less than 0.25 per cent), and it is interesting to consider what would happen if a larger clutch were laid. One cannot hope to observe a natural change (presumably a genetical mutation) which results in enough swifts laying clutches of four eggs instead of three to provide significant samples. However, by transferring young at hatching it is possible to compare the survival of young from broods of four with that from broods of two and three.

Swifts feed exclusively on airborne arthropods, the availability of which is greatly affected by the weather. In cold, wet summers the arthropods are less active and therefore less available to the swifts than in fine weather. At Oxford, Lack showed that in fine summers the average number of young produced per brood was highest from broods of three whereas in cold, wet summers it was highest from broods of two.

In the summers of 1958–61 inclusive I increased some broods of swift to four young by adding a newly hatched chick at the time that a fourth egg would have been expected to hatch. Subsequent survival is summarized in Table 1. In the summers of 1958, 1960 and 1961, the weather, and therefore the feeding conditions for swifts, were fairly good, and in 1959 they were exceptionally so. (Following the method used by Lack, the mean maximum temperatures during the nestling period were 68, 70, 72, and 70° F for the four years, respectively.)

It will be seen that, in all four years, the nestling mortality was markedly higher in larger broods, while in none of them did broods of four produce, on average, more surviving young per brood than broods of three. Even in 1959 when no nestling died in broods of 1, 2 or 3, 5 died out of 16 in broods of 4. (In three supplemented broods of four studied earlier by Lack, eight young flew, but four of the six in the exceptionally fine summer of 1957 apparently fledged prematurely and there is doubt as to whether any of these survived.)

Hence, even in an unusually fine summer, swifts starting with 4 young raised less per brood than those with broods of three, so that natural selection must operate against any laying of

TABLE 1

Year	Brood size	No. of broods	No. of young	No. lost	% lost	No. fledged brood
1958	1	7	7	2	28.6	0.71
	2	21	42	2	4.7	1.95
	3	4	12	1	8.3	2.75
	4	2	8	4*	50.0	2.00
1959	1	10	10	0	0	1.00
	2	15	30	0	0	2.00
	3	4	12	0	0	3.00
	4	4	16	5	31.2	2.75
1960	1	6	6	0	0	1.00
	2	18	36	2	5.6	1.89
	3	6	18	4	22.2	2.33
	4	5	20	14	70.0	1.20
1961	1	7	7	0	0	1.00
	2	18	36	1	2.8	1.95
	3	6	18	4	22.2	2.33
	4	5	20	13	65.0	1.40

* One more apparently left the nest prematurely.

clutches of four. As mentioned earlier, in some summers at Oxford broods of three have been the most productive and, in others, broods of two. Thus the present situation (of some individuals laying three eggs and others two, but virtually none four) is precisely what one would expect on the basis of natural selection. This conclusion is in agreement with the results for the other species cited, in all of which broods of larger than average size gave rise to fewer surviving young per brood than those of average size. This conclusion contrasts with the theory of Wynne-Edwards that animals might maintain their reproductive rate below the maximum possible in order to prevent overexploitation of the food supply, a theory which has yet to be demonstrated for any animal.

A Rejoinder to Perrins

V. C. WYNNE-EDWARDS

I doubt if it is possible from Dr. Perrins's results to convince the sceptic that swifts are "producing as many surviving young as possible," and to demonstrate in this way that natural selection has, as he believes, forced their fecundity up to the potential limit of efficiency. A second look at the figures soon puts the conclusion in doubt. I agree that in England clutches of four tend to be inefficient; they constitute a relatively infrequent group, although they seem to have been commoner in 1958–61 than they were in the 1946–56 period. Setting them aside we cannot be far wrong in assuming (having combined Perrins's samples for 1958–61 with those of Lack and Lack for 1946–51) that the remaining swifts breeding in the Oxford neighborhood in recent years have produced clutches of one, two, or three chicks in a percentage ratio of about 24: 61: 15. Two-chick broods have evidently been by far the most common.

The average productivity of each of these brood-sizes, in terms of flying young produced, is shown in Table 1.

From *Nature*, 201 (1964), 1148.

It is clear that the most productive brood-size over the 15 years in question has been three; by comparison with these, the broods of two have averaged fewer recruits and broods of one less than half as many. Dr. Perrins directs attention to the fact that in summers with poor weather the broods of three have no advantage over those of two: in fact, in the worst years like 1946–48 and 1953 they were fractionally (10 per cent) less productive. But in average and good years they yield a favorable margin, and, as Table 1 shows, over-all they have averaged not far short of 20 per cent more flying young than the broods of two. Why, one is entitled to ask, has natural selection not stabilized three as the optimum clutch, instead of just less than two, which is the observed average figure? Unless some additional hypothetical factor is postulated, like a differential survival of juvenile birds between leaving the nest and becoming adult breeders, Dr. Perrins's conclusion seems difficult to defend.

TABLE 1: *Young Raised per Brood*

Brood size	Lack and Lack	Lack	Perrins (preceding communication)
1	0.83	0.94	0.93
2	1.66	1.64	1.93
3	1.68	2.16	2.55

In Switzerland it is known that broods of three are still more productive (more than 90 per cent of the chicks so hatched grow into flying young); yet still only 67 per cent of the swifts there lay three eggs. On the hypothesis that selection at the individual level necessarily promotes maximum fecundity it appears *a fortiori* "curious that all the pairs there do not lay three eggs."

It is instructive to compare the swift with a bird like the red grouse (*Lagopus lagopus scoticus*), in which clutch-size is much

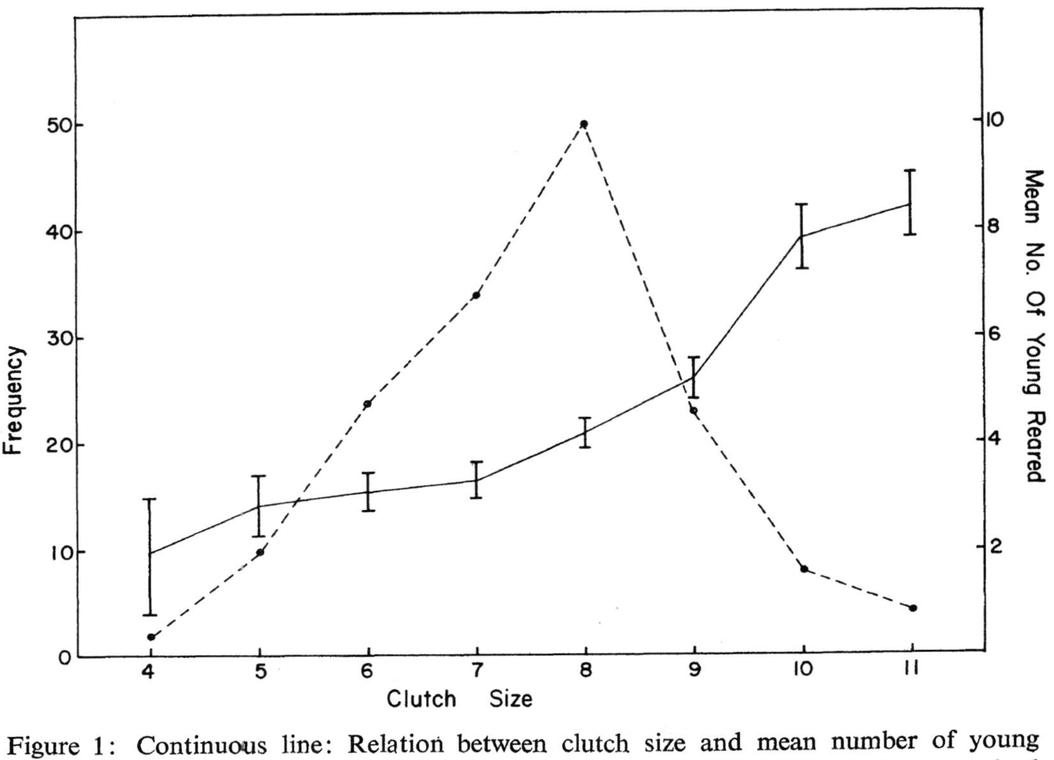

Figure 1: Continuous line: Relation between clutch size and mean number of young reared, in 153 broods of red grouse followed until August 12 (vertical lines show standard errors of means above and below). Broken line: Frequency distribution of clutch sizes in the same sample.

more variable, and production of flying young progressively increases with brood-size throughout the whole size range (Figure 1). Notwithstanding the fact that female grouse in suitable circumstances lay ten or more eggs and the more they lay the more young they produce, the mean clutch-size in north-east Scotland has been found to be only 7.5 eggs. (Reproductive performance is subject to annual fluctuations and, in fact, the average clutch has varied between 5.9 and 8.2 in different years since 1957.) Dr. Perrins's hypothesis would not fit this species any better than it does the swift; indeed, the situation here seems further still from being "precisely what would be expected on the basis of natural [inter-individual] selection."

On my alternative hypothesis that intergroup selection has fixed the fecundity range around the optimum level for recruitment to the population as a whole, the facts for both species seem reasonably intelligible.

III

Higher Levels of Organization

A small number of biologists—but some distinguished ones, such as W. C. Allee, Alfred E. Emerson, and Gordon A. Riley—have recognized from time to time a functional organization well beyond the level of the population or species. If, for instance, ecological communities are so organized, they must owe their properties at least partly to the natural selection of alternative communities. M. J. Dunbar's paper is perhaps the most thoughtful of recent contributions in this area.

5: The Evolution of Stability in Marine Environments: Natural Selection at the Level of the Ecosystem

M. J. DUNBAR

One of the most striking contrasts between the lower and the higher latitudes is manifested by the stability of the warm-adapted floras and faunas and the instability of the ecosystems of the cooler parts of the world. The argument developed here is that this contrast suggests, among other things, that selection may apply at the level of the ecosystem as well as at the levels of the individual and the specific population. Ecosystems can compete, and evolution of the stable ecosystem can be looked upon as a process of learning, analogous to the learning of regulated behavior in the nervous systems of animals.

Oscillations in Number

In this discussion, I am starting from the premise that oscillations are bad for any system and that violent oscillations are

From *The American Naturalist*, 94 (1960), 129–136.

often lethal. The more violent the oscillation in specific numbers in any ecological situation, the greater the danger of extinction of species, at least of local extinction, causing serious disturbance of the community, possibly the extinction of the whole system, again locally.

Population oscillations of fairly wide amplitude are well known in terrestrial environments among rodents and their predators, game birds, rabbits, and insects, and much study has been devoted to them. Oscillations of this sort, in which a primary oscillator, a herbivore, causes sympathetic oscillations in carnivore populations which in turn take part (however small a part) in reducing the amplitude of the oscillations of the herbivore, appear only in relatively simple ecosystems, containing a limited, usually quite small, number of species. Such simplicity is found as a rule only in cool climates with marked seasonal variation, that is to say, in temperate and polar climates. (A special case might be made here for desert regions, in which ecosystems are also simple.) Oscillations of this sort are absent in tropical and subtropical environments, which foster much more complex ecosystems in which there is great multiplicity of energy paths along which overloadings can be released, with consequent great decrease or virtual elimination of the time-lag effect which is largely responsible for the oscillations in the simpler systems.

The simplicity of the ecosystems in cool and cold climates is presumably partly or mainly a result of the novelty of the present polar conditions. Whether or not we accept the views of Zenkevitch on the ancient origin of the cold-adapted fauna or the more conservative estimates of the age of the present polar cooling of Barghoorn and Durham, there can be little doubt that the conditions existing during the Pleistocene were special. The short million years since the end of the Pliocene, or the few million years of glacial climate possibly since some time in the Pliocene or Miocene, and the much shorter time available for colonization of areas actually glaciated, have not been long enough to allow the adaptation to the new conditions toward the poles of more than a relatively small number of species. Such adapted forms have by establishing themselves first, further lowered the chances of successful establishment by others, the more so because the number of possible niches and habitats in the colder regions is very much less than in warmer parts of the world.

Seasonal Oscillations

What is true of fluctuations whose periods occupy a few years is also true of *seasonal* oscillations—they are most marked in temperate and polar regions, and absent or poorly developed in the tropics both on land and in the sea. In the plankton, for instance, upon whose numerical behavior everything else in the sea depends, production is continuous at a steady controlled level all year round in tropical waters, according to Steemann Nielsen. Plant growth on land, in equatorial regions, has a similar year-round activity; individual species may develop periodicities of flowering and of new leaf growth, not necessarily annual periodicities, but the total mass and the total activity level remain fairly constant. Steemann Nielsen describes extraordinary stability in the planktonic system in tropical seas. The phytoplankton does not at any time completely consume the available phosphate and nitrate supplies, which are continuously present at low levels; the zooplankton does not deplete the phytoplankton, and predation upon the zooplankton does not, apparently, cause oscillations. This means that the levels of production are delicately controlled in a hydrographic system which is itself highly stable. Except for certain areas of large-scale upwelling, which are among the most productive regions in the world, the warmer seas are very stable, and vertical movements of water are effectively and stubbornly resisted. The supply of nutrient salts is thus limited to (1) regeneration within the euphotic layer, and (2) such quantities as can enter any given area from outside by horizontal transport. These two sources have to maintain the equilibrium against constant loss from the system of salts contained in detritus which sinks out of the photosynthetic layer before it is mineralized. The quantities involved in this loss are not yet known.

The contrast between this tropical stability and the sharply oscillating annual cycles of standing crop in temperate and polar systems is obvious, and it is equally clear that the greatly differing climatic conditions are intimately associated with the contrast. The annual variation of light intensity and angle of incidence, the larger amplitude of annual temperature variation, and the unstable condition of the colder waters in winter, are all involved here. It may not be so obvious, however, that these annual oscillations in both marine and terrestrial ecosystems have something important in common with the "fur-bearer" type of oscillations already mentioned. Both types can be attributed to climate; the former as just described and the latter owing to the simplicity of the ecosystem, which is itself a result of climatic demands. Both groups of oscillations are to a high degree the result of the onset of the glacial climate. On the original premise, then, that oscillations are disadvantageous both to the individual species and to the ecosystem as a whole, I conclude that the steady systems of the tropics are the result of long evolution and that oscillations observed in the higher latitudes are systems of nonadaptation. Arctic and antarctic faunas are immature, still in the elementary stages of evolutionary "learning."

Adaptations to Cold Climates

The next thing to look for are signs of incipient adaptation, in this ecological sense, to life in cold climates. Adaptation in this case would be a matter of changes in the system which would tend to (1) increase the complexity of the ecosystem, that is, increase the number of species involved; (2) lower the rates of production from the maxima possible in both plants and animals; (3) spread the rate of grazing upon the plants as evenly as possible over the 12 months of the year, and (4) carry these processes up to higher trophic levels. This would involve control of the breeding rates of both plant and animal populations, and possibly also the regulation of the manner in which individuals in the herbivore and carnivore populations grow, and of their energy requirements. There is evidence that processes of this sort have in fact been evolved, or are in process of evolution.

1. The findings of Thorson can profitably be interpreted in this light. Thorson found that in marine shallow water benthos in temperate and arctic regions, the pelagic larva may or may not be retained. If it is retained, the spawning is restricted to a short period in the spring calculated to coincide with the abundance of the spring plankton; if it is not, the spawning period is much longer and could extend throughout the year. There is a direct relationship between the mean temperature of the environment and the proportion of species which retain the pelagic larva. The stabilizing effect of the loss of larva and production of larger eggs is thus best developed in the colder regions.

2. Steele, working on the eastern Canadian arctic shallow water Amphipoda, concludes that whereas the littoral species have developed short seasonal breeding periods in the spring or summer, the benthonic forms as a rule breed all the year round. The immediate or proximate reason for this may well be that the detrital or bacterial food of these species is present the year round anyway; nevertheless, the ecological effect is toward high biomass in winter.

3. Recent work on the marine fauna of the arctic regions has brought to light the fact that many species produce their young not simply at any time of year, but specifically in the dead of winter. In a situation in which the spring bloom of phytoplankton is sudden and violent, in which there may be no fall bloom at all and the winter stock of plant food tends to be quite small, so far as we know, one would have expected that all the animal species which depended on phytoplankton in the younger stages would have evolved a breeding cycle that would release the young at the most propitious moment, at some time between the beginning and the peak of the spring phytoplankton

bloom. Many species of course do this, notably the copepods, but there are many that do not, and as the study of arctic breeding cycles advances, more such species appear. MacGinitie gives many examples of invertebrate species in the waters of the Point Barrow area, Alaska, which begin to develop their eggs in October and later, and many which produce ripe eggs at that time; some of the amphipods had hatched young in the marsupium. Dunbar records the dominant arctic pelagic amphipod *Themisto libellula* as maturing late in the autumn or in the early winter and carrying hatched young in December. *Gammarus setosus* breeds in the colder part of the year, and differs in this respect from its close relative *G. oceanicus.* And among the 114 species of amphipods taken in Ungava, making up a very large material collected over four seasons, only 55 species were found to include ovigerous or paedigerous females during the spring and summer months, from June to September. The material collected by the Russian drifting station expedition in the polar basin in 1950–1951 included several species recorded as spawning in winter. This winter breeding has the effect of increasing the number of species in the system. This is analogous to the adoption of nocturnal or diurnal habit, and it also spreads the rate of predation on the available food more evenly over the year.

4. There is something possibly significant in this regard in the large size and lowered growth rate of arctic poikilotherms. Harvey, in discussing growth and metabolism in marine poikilotherms, writes: "In addition to this general inverse relation between age or size of animals and losses by respiration, there is an inverse relation between age and growth rate (the daily percentage increase in organic matter). It appears usual that growth rate decreases more rapidly with increasing size than respiration decreases. . . . In consequence, a greater proportion of food assimilated by young animals is built into new tissue than by old animals. Hence the same rate of plant production may permit a greater biomass of a stable community consisting mostly of aged, larger, slow-growing, slow-respiring animals than of one mainly composed of small quick-growing animals.

The latter fauna, however, may synthesise more animal tissue yearly, the rate of turnover of living tissue in the animals being greater."

This, the development of large size and slow growth and metabolic rates, is precisely what happens in cold polar waters, which raises the question whether this situation may not be the result of adaptation to the total ecological condition, assuming that there is selection favoring stability.

5. The two-phase or alternating breeding cycle demonstrated in certain cold-water members of the zooplankton, such as *Themisto libellula*, *Sagitta elegans*, *Thysanoessa* spp., may also be interpreted in this light, since it leads to stability of the species population and maintenance of large numbers of individuals under conditions of slow growth, slow maturation, and probably limited food supply. In this type of cycle two broods coexist in the same body of water, the two broods being separated in time so that they may be partially or completely isolated from each other reproductively.

6. Finally, there is an example of stability which is as striking as that of the tropical plankton, and about which a little more is known. The families of oceanic birds are remarkable for the stability of their population numbers. The range of a species may increase or decrease over a period of years, as in the case of the Fulmar Petrel in the British Isles or of the Gannet in the North Atlantic, but oscillations of the sort discussed here do not appear to occur. They breed on small island groups or rock cliff shores, with a food supply that must be considered to be virtually without limit for practical purposes, and with great mobility. In spite of this plenty, the oceanic birds are most modest in their breeding rates. The numbers of eggs laid are small, often only one per pair, and breeding may not occur every year per individual, at least in some species. Why this frugality amid such plenty? I suggest that these are highly evolved stable populations which have in the past been subjected to the stress of oscillation in an oscillating system, and that they have responded to selection for this self-regulating character of a restricted breeding rate, tending toward stability. Baker

quotes examples of many tropical (terrestrial) birds which lay only two eggs in a clutch, though they belong to genera in which five or six eggs are the rule among temperate representatives.

Some of the sea-bird populations mentioned above breed in polar or temperate regions. But if they do, they are migrants, and are present toward the poles only when the food supply is maximal for the year. As such they serve to cut down the animal marine populations when the latter are high, and depart for other parts when the food supply grows less; they do not prey upon animals during the low-level period in the oscillation. Moreover, birds, as homotherms, are less affected physiologically by the polar climates than are the poikilotherms—the problems of adaptation are much simpler. This is true also of the sea mammals, whose numbers also appear to be steady. Certain of the most numerous sea mammals that breed in the north are migratory; the phenomenon of migration itself may perhaps profitably be considered as the result of evolution at the ecological or ecosystem-level, as something which tends to stabilize both the systems of the breeding areas and those of the wintering areas.

The case made here, then, is that the stable nonoscillating system is the "ideal" and the product of a long process of evolution; and that the oscillating systems of regions affected by the present glacial climate of the world are in process of evolution toward better adjustment.

Selection of Ecosystems

As to the mechanisms by which selection might take effect at this level, they are of the ordinary Darwinian sort except that the criterion for selection is survival of the system rather than of the individual or even the species. For instance, suppose an ecosystem, locally defined, begins to develop oscillations to a lethal degree, a degree such that one or more vital parts are not able to survive; the resulting empty environmental space, as in Cuvierian cataclysms, is available for occupation by communities from the adjacent regions; and these adjacent systems, as their survival suggests, are not of precisely the same constitution as the extinguished system. Perhaps a drift effect has occurred in certain species, or some other isolation effect. One or more of the specific elements will have growth rates, breeding potential and/or metabolic adjustment to temperature different from the former system; and if the difference is favorable to the continued survival of the system, its chances of survival are enhanced. In this way the system dominant in any geographic region changes, and changes (if the present assumptions are correct) in the direction of greater stability.

In this sort of selection, characters which may be of immediate advantage to the species are not necessarily selected into the stock in the long run, as may happen to the character which may be advantageous to the individual but ultimately lethal to the species. A breeding rate high enough to give one species an immediate advantage over another may ultimately prove to be lethal to the whole system. It thus becomes possible, in fact necessary, for selection to favor *slower* growth rate, *longer* time to reach maturity, *winter* spawning, and so on, under certain determining circumstances. It should be pointed out here that the model suggested in this paper has already been called for, as it were, by Hutchinson who writes: "So far little attention has been paid to the problem of changes in the properties of populations of the greatest demographic interest.... A more systematic study of evolutionary change in fecundity, mean life span, age and duration of reproductive activity and length of post-reproductive life is clearly needed. The most interesting models that might be devised would be those in which selection operated in favor of low fecundity, long pre-reproductive life and on any aspect of post-reproductive life." There is much more in the paper just quoted of direct relevance to the thesis put forward here. Slobodkin, also, has pointed out that in certain circumstances a low rate of reproduction may have a greater selective advantage than a high rate. Utida comes to the conclusion from the mathematical approach that selection in favor of dampening of oscillations is to be expected, and points out that both Hutchinson and Slobodkin had come to similar conclusions from other directions. Chitty describes a noninfectious pathological condition which develops in *Microtus agrestis* under stress of crowding, and adds: "The main theoretical difficulty is to explain why such a deleterious condition, if controlled by an hereditary factor, has not been eliminated by natural selection." Finally, and in the contrary sense, Barnes, describing the synchronization of spawning in *Balanus balanoides* with the spring phytoplankton outburst, regards such synchronization in arctic regions as essential; it is in fact essential only within the framework of classical selection theory, and it is clear that there are many forms in which no apparent effort at such synchronization is made.

There are probably strict limits to the degree to which an ecosystem can stabilize itself in an environment as variable (seasonally) as the climates of the higher latitudes. Perhaps stability comparable to the tropical stability was not achieved in higher latitudes even before the onset of the glacial climates. The thesis here is simply that there is a selection acting constantly in favor of greater stability, whether or not such selection meets a situation of equilibrium with an oscillating environment, beyond which greater stability of the system may become impossible.

Summary

Starting from the premise that oscillations are dangerous for any system and that violent oscillations may be lethal, this paper contrasts the highly stable production systems of tropical waters with the seasonal and longer-term oscillations of temperate and polar waters. The differences are climatically determined, and since the present glacial type of climate is young in the climatic history of the earth, the ecological systems of the higher latitudes are considered as immature and at a low level of adaptation. That they may be in process of evolution toward greater stability is suggested by a number of phenomena, such as the development of large, slow-respiring, slow-growing individuals, and the production of the young in many arctic invertebrates in mid-winter or late fall. These and other observed peculiarities of high latitude fauna tend to make the most efficient use of the available plant food and to spread the cropping pressure over as much of the year as possible. Oceanic birds are cited as examples in which stable populations have been achieved by evolution of lower breeding rates, and the phosphate and nitrate cycles in the upper layers of tropical seas are discussed. It is emphasized that selection here is operating at the level of the ecosystem; competition is between systems rather than between individuals or specific populations.

IV

Sex Ratio

It is remarkable that until 1930 the biological significance of the relative numbers of the sexes was almost totally ignored. Then R. A. Fisher treated the subject briefly but cogently in his *Genetical Theory of Natural Selection*. The subject was then ignored for more than twenty years. Since the early 1950s sex ratio has been a live field of theoretical inquiry, and the theoretical models are now being applied by ecologists studying natural populations.

Two schools of theorists are readily distinguished. A minority look upon sex ratio as an adaptation at the population level, and they speak of such things as the optimum sex ratio for the population. Most of the theorists follow Fisher and regard sex ratio as merely the statistical summation of effects of selection at the individual level. They may speak of an optimum sex ratio of a litter or clutch, but never of the population. These two approaches are nicely illustrated in the following papers, and their divergent points of view are explicitly contrasted in the final note by Edwards.

6 : *Natural Selection and the Sex Ratio*

W. F. BODMER

A. W. F. EDWARDS

Darwin considered whether the sex ratio is subject to Natural Selection. After surveying the available statistics and investigating the selective effects of infanticide at some length he concludes by saying that as far as he can see there is no selective advantage attached to a particular sex ratio, and that "I now see that the whole problem is so intricate that it is safer to leave its solution to the future."

There the matter lay until it was taken up by Fisher, who outlined the solution to the problem. His approach depends on essentially economic arguments which must take into account the parental expenditure of effort involved in the rearing of offspring to maturity. This expenditure will be a function of the time and energy which the parents are induced to spend in rearing their offspring, and must therefore depend on the mortality during the period of parental expenditure. Here is a shortened account of Fisher's argument:

From *Annals of Human Genetics*, 24 (1960), 239–244. Reprinted by permission of Cambridge University Press.

Let us consider the reproductive value, or the relative genetic contribution to future generations, of some offspring at the moment when parental expenditure of effort on them has just ceased. It is clear that the total reproductive value of males in a generation of offspring is equal to the total value of the females, because each sex must supply half the ancestry of future generations. From this it follows that the sex ratio will so adjust itself, under the influence of Natural Selection, that the total parental expenditure of effort incurred in respect of children of each sex shall be equal; for if this were not so and the total expenditure incurred in producing males, for instance, were less than that incurred in producing females, then parents genetically inclined to producing males in excess would, for the same expenditure, produce a greater amount of reproductive value, with the result that in future generations more males would be produced. Selection would thus raise the sex ratio until the expenditure upon males became equal to that upon females.

In 1953, Shaw and Mohler tackled the problem, but they consider that "Fisher's treatment is phrased in non-genetical terms and does not lend itself to further development." They therefore "ignore instances involving parental care" and come to the anticipated conclusion that the equilibrium sex ratio at conception should be one-half (we define the sex ratio to be the proportion of males). Since the production of offspring by a sexual organism always involves expenditure by the parents, their treatment is necessarily incomplete.

Development of Fisher's Theory

The purpose of this paper is to put Fisher's theory, and some of the conclusions that may be drawn from it, on an analytic basis. Let us consider first the selective advantage attached to reproduction with a given primary, or conception, sex ratio. Suppose that the mean primary sex ratio in a population is X, and in a given part of the population is x. Let the proportion of males living until the end of the period of parental expenditure be M in the whole population and m in that part of the population producing with sex ratio x, and let the corresponding proportions of females be F and f.

Now consider the reproductive value of an individual from the "x" part of the population at the moment when parental expenditure on his or her behalf has just ceased. The reproductive value of a male is inversely proportional to the number of males at that stage in the whole population, or proportional to $1/XM$. Similarly, the reproductive value of a female is proportional to $1/(1 - X)F$. Thus the average reproductive value of an individual reproducing with sex ratio x is given by the mean of these values weighted by the proportions of the two sexes living at the end of the period of parental expenditure, and is proportional to

$$\frac{\dfrac{xm}{XM} + \dfrac{(1-x)f}{(1-X)F}}{xm + (1-x)f}.$$

If we now write $xm/[xm + (1 - x)f] = x'$, the sex ratio at the end of the period of parental expenditure, and $XM/[XM + (1 - X)F] = X'$, then the reproductive value is proportional to $x'(1 - X') + (1 - x')X'$.

We now find the expression for the average parental expenditure required to raise one child to the end of the period of parental expenditure in the "x" part of the population. Suppose that the expected total expenditure on a child is proportional to the probability of its surviving to the end of the period. This assumes that children dying before the end of the period incur a negligible expenditure compared with those who survive. This is likely in man at least, since the majority of deaths occur early in pregnancy, at a time when the parental expenditure is small. Let the expected expenditure on a male child be to that on a female child as h to $1 - h$. Thus the expected expenditure on a male child is mh and on a female $f(1 - h)$. The average expenditure per child conceived is therefore $xmh + (1 - x)f(1 - h)$, and, dividing this by the probability of a child living to the end of the period of parental expenditure, we have the average expenditure required to raise one child to the end of that period:

$$\frac{xmh + (1-x)f(1-h)}{xm + (1-x)f}\ .$$

Substituting for x' as before gives $x'h + (1 - x')(1 - h)$.

The reproductive value per unit parental expenditure in the "x" part of the population is therefore proportional to

$$R = \frac{x'(1-X') + (1-x')X'}{x'h + (1-x')(1-h)}\ ,$$

and this expression is a measure of the selective advantage attached to reproduction with particular sex and parental expenditure ratios.

The value of R will be a maximum for variation in x' when

$$\frac{dR}{dx'} = \frac{1 - h - X'}{(x'h + (1-x')(1-h))^2} = 0.$$

Thus when $X' = 1 - h$ the selective advantage is a maximum, and is seen to be independent of x', so that all sex ratios are then equally advantageous. If X' is less than $1 - h$ large values of x' are at an advantage and X' moves toward $1 - h$ in the next generation; if X' is greater than $1 - h$ small values are at an advantage, and X' again moves toward $1 - h$. The population is therefore at a stable equilibrium when the mean sex ratio at the end of the period of parental expenditure, X', is $1 - h$. We then have $X'h = (1 - X')(1 - h)$, which gives Fisher's Law, that at equilibrium the total parental expenditure incurred in respect of children of each sex is equal.

It is important to note that this equilibrium refers to the sex ratio X' at the end of the period of parental expenditure, and not to the primary sex ratio or the sex ratio at birth. The selective advantage R also depends only on the sex ratios x' and X', so that allowing for variation in x' takes into account both variation in the primary sex ratio and in the differential mortality during the period of parental dependence. For most species it is likely that the major part of any differences between the two sexes in parental expenditure will depend on this differential mortality and not on any differential demands which the young make on their parents. The quantity h, which is a measure of the latter more restricted part of the differential expenditure, may therefore be expected to remain constant and near to the value one-half. We have thus shown that Natural Selection will tend to maintain the sex ratio at the end of the period of parental expenditure near the value one-half. The numerical equality of the sexes in man in the age group 15–20 years has frequently been commented upon, but in spite of Fisher's theory it has usually been assumed that the optimum sex ratio is one-half at the reproductive age because the chance of encounter between the sexes is then a maximum. However, it will not be surprising if modern human populations are not in equilibrium for the sex ratio, because prenatal and infant mortalities have been changing very quickly in recent years, and differential mortalities may have been changing more quickly then selection can change them or the primary sex ratio.

Progress of a Population Toward Equilibrium

As an example of the implications of the above analysis in the study of changes in the sex ratio, we take the case of a population in which $h = \frac{1}{2}$, and suppose it to be slightly displaced from its equilibrium position, which, since $h = \frac{1}{2}$, will be at $X' = \frac{1}{2}$. Dropping the primes from our notation, let the probability density function of the sex ratio at the end of the period of parental expenditure in generation 0 be $f(x)$, where

$$\int_0^1 f(x)dx = 1,$$

and let the distribution have mean X_0 and variance V_0. The selective advantage of a sex ratio x in a population with mean ratio X_0 is proportional to $x(1 - X_0) + (1 - x)X_0$, since $h = \frac{1}{2}$, or to $X_0 + x(1 - 2X_0)$.

In deriving the distribution of sex ratios in the next generation we shall, for simplicity, assume that we can multiply the probability density at a given sex ratio by the relevant selective advantage. One genetic system for which this is true is where there is complete genetic determination of the ability of one particular sex to reproduce with a given sex ratio. We may therefore expect that the rate of progress toward equilibrium that is obtained will be an upper limit to the range of possible rates that would be obtained under more realistic, and complex, assumptions.

Performing this multiplication we obtain the probability density function for the next generation:

$$k[X_0 f(x) + xf(x)(1-2X_0)],$$

where k is chosen so that the population size remains unity. Integrating the distribution in generation 1 we find

$$1/k = X_0 + (1-2X_0)X_0,$$

since

$$\int_0^1 xf(x)dx = X_0.$$

The mean of generation 1 is therefore

$$X_1 = \frac{X_0 \int_0^1 xf(x)dx + (1-2X_0) \int_0^1 x^2 f(x)dx}{X_0 + (1-2X_0)X_0}$$

$$= \frac{X_0^2 + (1-2X_0)(V_0 + X_0^2)}{2X_0(1-X_0)}$$

$$= X_0 + \frac{V_0}{2}\left(\frac{1}{X_0} - \frac{1}{1-X_0}\right).$$

Thus the change in the mean is

$$X_1 - X_0 = \frac{V_0}{2}\left(\frac{1}{X_0} - \frac{1}{1 - X_0}\right).$$

Similarly, the variance of generation 1 is

$$V_1 = \frac{X_0\int_0^1 x^2 f(x)dx + (1-2X_0)\int_0^1 x^3 f(x)dx}{X_0 + (1-2X_0)X_0} - X_1^2.$$

Now

$$\int_0^1 x^3 f(x)dx = \int_0^1 (x-X_0)^3 f(x)dx + 3X_0\int_0^1 x^2 f(x)dx$$

$$- 3X_0^2\int_0^1 xf(x)dx + X_0^3$$

$$= 3X_0(V_0 + X_0^2) - 3X_0^3 + X_0^3$$

$$= 3X_0V_0 + X_0^3,$$

neglecting the third moment about the mean of the original distribution, which in any case is zero if the distribution is symmetrical.

Thus

$$V_1 = \frac{X_0(V_0 + X_0^2) + (1-2X_0)X_0(3V_0 + X_0^2)}{2X_0(1-X_0)}$$

$$- \frac{[X_0^2 + (1-2X_0)(V_0 + X_0^2)]^2}{4X_0^2(1-X_0)^2}$$

$$= V_0 - \left[\frac{V_0}{2}\left(\frac{1}{X_0} - \frac{1}{1-X_0}\right)\right].$$

We already have

$$X_1 = X_0 + \frac{V_0}{2}\left(\frac{1}{X_0} - \frac{1}{1-X_0}\right)$$

so that

$$V_0 - V_1 = (X_0 - X_1)^2,$$

and there is a decrease in the variance from generation to generation.

If the population is near the equilibrium position

$$\frac{V_0}{2}\left(\frac{1}{X_0} - \frac{1}{1 - X_0}\right)$$

is small, and its square may be neglected, in which case $V_1 = V_0$. Thus near the equilibrium we assume that the variance is constant, and the change in the mean from the nth generation to the $(n + 1)$th is given by

$$X_{n+1} - X_n = \frac{V}{2}\left(\frac{1}{X_n} - \frac{1}{1 - X_n}\right).$$

If $X_n - \frac{1}{2}$ is small, as it will be near the equilibrium, a linear approximation to the above equation is

$$X_{n+1} - \tfrac{1}{2} = (X_n - \tfrac{1}{2})(1 - 4V),$$

with error of order $(X_n - \frac{1}{2})^3$. Writing $X - \frac{1}{2} = Y$, the difference between the sex ratio and its equilibrium value,

$$Y_n = Y_0(1 - 4V)^n.$$

From these results we can see that *the rate of approach of a population's sex ratio to an equilibrium value of one-half is directly proportional to the genetic variance in sex ratio of that population.* In populations where the equilibrium value is nearly one-half we may expect this result to be at least a good approximation.

Discussion

Knowledge of the variances in sex ratio of populations is very limited. It applies exclusively to the sex ratio at birth, whereas from our analysis we see that knowledge of the variance of the sex ratio at the end of the period of parental expenditure would be more useful. For man, Edwards obtained an estimate of the total variance of the sex ratio at birth of 0.0025, and most of this is probably environmental. An alternative approximate expression for Y_n is

$$Y_n = Y_0 e^{-4V_n}$$

from which we see that the deviation of the sex ratio from one-half will decrease by a factor $e = 2.718$ in $1/4V$ generations. For example, to reduce the sex ratio from 0.5200 to 0.5074 will take this time. If the genetic variance is 0.0025 this is 100 generations, or about 2000 years for man. If, however, the variance is much less, as it may well be, the time is proportionately longer. Thus changes in the human sex ratio due to Natural Selection are almost certainly too slow to be detected over the period for which data are available, although they may not be slow in comparison with many other evolutionary changes.

Since the rate of approach to equilibrium depends on the variance, large variances should be selected for if the ability of a population to approach equilibrium rapidly is advantageous. Now our analysis has only considered intrapopulation selection and is in no way relevant to interpopulation selection. In most sexually reproducing species the reproductive potential of a population will be limited by the number of females it contains. Hence for interpopulation selection an excess of females may be an advantage, although within any population the changes in the sex ratio will be as analyzed above. There would thus be no advantage in having a large variance and consequently the ability to approach equilibrium rapidly. This is borne out by the small variances in the sex ratio that have in fact been observed.

It should be possible to mimic the effect of Natural Selection on the sex ratio by artificially changing the differential infant mortality. For example, if in a mammalian population half the males are killed at birth, selection should act so as to bring back the sex ratio at the end of the period of parental expenditure to its original value. The main difficulty in any form of selection for the sex ratio is that the estimate of the probability of a birth being male in a single family, on which selection depends, has a very large variance. A further difficulty, unforeseen by previous experimentalists, is that the type of selection that we have been considering may exert a stronger selection pressure toward equilibrium than the experimental design can exert away from equilibrium. In this respect the above-mentioned experiment, which has been started in this department, is at an advantage. In practice the small amount of variance available will probably only allow changes that are too small to be detected in a reasonable time. It is also likely that in selecting for a change we select for a combination of changed primary sex ratio and infant mortality, which together may have been stabilized by Natural Selection.

It is clear that models for the action of specific genes on the sex ratio must take into account the selective forces which we have analyzed. This may call for some modification of the models proposed by Shaw and of Bennett's discussion of Wallace's data on the equilibrium of the "sex ratio" gene in *Drosophila*.

For some time it has been assumed that in order to explain the apparent distribution of the sex ratio with age in human populations it is necessary to postulate that the primary sex ratio is high. We have shown that the prevailing sex ratio has arisen through the interaction of the primary sex ratio and the mortality rates for males and females, and the existing situation in man can be explained equally well by a high primary sex ratio and considerable sex-differential mortality, or by a primary sex ratio near to one-half and little differential mortality. The direct evidence for a high primary sex ratio seems to be somewhat contradictory and has been critically reviewed by McKeown and Lowe. They state that the existing data are inadequate to justify any assumption about its value; that "at least half of all abortions occur in the first three months" of pregnancy, and that the sex ratio of live foetuses is about one-half at the seventh month. If the primary sex ratio is high there must therefore be considerable differential mortality during early pregnancy, and, since the expenditure on offspring dying during this period is probably small, the difference in expenditure between males and females dying will also be small, and will be even less if the primary sex ratio is near to one-half. It therefore seems likely that the major difference in expenditure between males and females dying during the period of parental expenditure will be incurred in respect of offspring dying just before, and in the year following, birth, when there is considerable differential mortality. Hence this is probably the major contribution to differential expenditure, apart from differences on individual males and females for which we have allowed in the parameter h, which we are neglecting in assuming that the expenditure on an individual conceived is proportional to the probability of its living to the end of the period of parental expenditure. However, it seems unlikely that taking this expenditure into account will alter our basic conclusion that the sex ratio at the end of the period of parental expenditure will be stabilized at a value not far removed from one half.

Summary

We have put Fisher's theory of the control of the sex ratio by Natural Selection on an analytic basis. This has enabled us to derive an expression for the selective advantage attached to reproduction with a given sex ratio, and to show that this depends on the sex ratio at the end of the period in which the offspring incur expenditure by their parents. It is this sex ratio which is probably stabilized near the value one-half by Natural Selection. The rate of approach of a population to its equilibrium sex ratio depends on the available genetic variance in the sex ratio, and since this is probably small, evolutionary changes in the sex ratio of natural populations will almost certainly be too slow to detect.

7: *Evolutionary Origin of Sexual Differentiation and the Sex Ratio*

H. KALMUS

C. A. B. SMITH

Studies of sex ratio usually start from the assumption that it is specifically advantageous for a species to divide itself into two sexes in equal numbers. They then proceed to special explanations why this ratio should break down or be imperfectly realized in particular instances. The purpose of this article is to investigate: (1) why and when a 1:1 sex ratio might be advantageous; (2) how it might have arisen.

Advantages of Sexual Differentiation

Experience suggests that inbreeding in a population is usually disadvantageous, and often markedly so. One of the prerequisites for outbreeding is cross-fertilization, and this also provides

From *Nature*, 186 (1960), 1004–1006.

a large number of different gene permutations through genetic recombination. Such genetical diversity can be important when there are abrupt changes in environmental conditions, and without it the species may not be able to survive. A possible further function of recombination is the replacement of deleterious mutants. Without such replacement, mutants which are only slightly disadvantageous could slowly accumulate by chance in small isolated populations, and in time the species might die out. These three advantages of cross-fertilization suggest why it should be biologically important.

Although it is possible to have cross-fertilization without sexual differentiation (a condition called isogamy), we show in what follows that it is greatly assisted by such differentiation. We suggest that this is the fundamental reason for the evolution of sex in a population, and why it should have arisen repeatedly in the history of a single species.

Fertilization also serves the propagation of a species. Self-fertilization is much simpler and a less-hazardous procedure than cross-fertilization. Hence, unless there is a really effective barrier to self-fertilization, it is likely to be very much more common than crossing, especially in a species which is rare and widely scattered, and this could result in a harmful degree of inbreeding.

There is accordingly a conflict in many species between the advantages of cross-fertilization on one hand, with its avoidance of inbreeding, and self-fertilization or clonal development on the other hand, with their much greater ease of propagation. Survival of the species may depend on some compromise.

Another pair of contradictory tendencies closely connected with our problem are the respective advantages of being large and relatively secure, or of being small, numerous, and mobile. In multicellular organisms compromise solutions are reached which combine both these tendencies, either by the sequence in time of large soma and small gametes or spores, or by the simultaneous production of large and small gametes in the same species.

Advantages of Numerical Sex-Equality

One aspect of such developments is the establishment of certain numerical equilibria, of which, as we shall show, the sex-ratio is an example. Let us start by considering a situation in which cross-fertilization is guaranteed by self-sterility. For example, in many unicellular species two gametes from the same clone cannot unite to form zygotes, but those from different sub-strains may be able to do so even when they are alike in outward appearance. If there are only two such sub-strains, suppose they occur with respective numbers n_1 and n_2: then the number of opportunities of such a mating is proportional to the number of pairs of individuals chosen one from the one strain and one from the other, and this number is n_1n_2. Now if the total number of individuals $(n_1 + n_2)$ is assumed to be effectively determined by environmental conditions, we (will) get maximum propagation when n_1n_2 has its maximum value for given $(n_1 + n_2)$. But:

$$n_1n_2 = \tfrac{1}{4}(n_1 + n_2)^2 - \tfrac{1}{4}(n_1 - n_2)^2$$

and hence n_1n_2 is greatest when $\tfrac{1}{4}\,(n_1 - n_2)^2$ is least, that is, when $n_1 - n_2 = 0$, and the two sub-strains occur in equal numbers. (Similarly, with any number of sub-strains which are cross-fertile but self-sterile, the optimal situation is that in which all strains occur in equal proportions.) Mechanisms which ensure numerical equality of the two kinds of gametes will therefore confer a selective advantage: and we may expect in such cases numerical equality of morphologically similar types of gametes. Situations of this kind are observed among other cases in green flagellates and also in the infusoria where nuclei are exchanged between two mating types.

The products of isogamy in algae and of conjugation in infusoria do not undergo long and complex developmental changes and therefore do not need much reserve substances. On the other hand, such substances occur more or less abundantly in the zygotes of many higher plants and animals. If such species also require a high degree of genetical recombination, and hence cross-fertilization, the maximum propagation will be achieved by providing the greatest possible number of zygotes with sufficient reserves. Kalmus has shown that under certain plausible conditions this will happen if there is an unequal distribution of reserve material to two different kinds ("small" and "large") of mobile aquatic gametes, but in such a way that the reserve as a whole is about equally divided between a large number of the small gametes and a small number of the large ones. In organisms having external fertilization, such as echinoderms, bivalves, fish and others, the total masses of spermatozoa and ova are indeed found to be approximately equal. A similar situation is the comparable bulk of pollen and ovarial material in some wind-pollinated plants.

In monoecious plants and hermaphroditic animals the proportion of sperm and ova (pollen and ovules) is regulated by physiological means in each individual, and questions of sex ratio do not arise. The problem is somewhat different in bisexual animals and plants where the proportion of sperms and ova depends on the ratio of males to females in the population. The assimilatory powers of males and females are not very different in many species, and each can produce about the same volume of sex products. Therefore the optimal situation of near equality of bulk of spermatozoa and ova can be safeguarded by near equality of the numbers of individuals in the two sexes. This occurs in noncopulating species of the following animal groups: Coelenterata, Scolecida, Mollusca, Brachiopoda, Enteropneusta, Echinodermata, Tunicata, Ascidia, Cyclostomata, Pisces, and Amphibia. In plants, comparable relations exist among Myxomycetes, Phycomycetes, as well as among the dioecious flowering plants.

In the case of animals with internal fertilization which have two sexes (copulating, gonochoristic forms) we shall require the chances of encounter of a male and female to be a maximum; we have already shown that this occurs when the two sexes have equal frequencies, assuming the total numbers to be limited by other factors.

Inequality in the numbers of the sexes can also be disadvantageous in another way. If the population contains, let us say, only a small number of females, then half the genes in the next generation must be derived from these females, however many males there may be. It follows that the genetical diversity of the population will thereby be restricted; in fact, there will be a large number of pairs of halfsibs in the next generation, and so presumably a large number of matings of related individuals: and this will lead to inbreeding. Sewall Wright has devised a measure of this restriction of genetical variation in a small population which he calls the "effective size" of the population. If there are n_1 males and n_2 females, and the population breeds at random, the effective size is:

$$N = \frac{4n_1n_2}{n_1 + n_2} = n_1 + n_2 - \frac{(n_1 - n_2)^2}{n_1 + n_2}$$

The greater the "effective size" the smaller the degree of inbreeding, other things being equal; hence the optimal situation occurs when the effective size N is a maximum. For a given total size $n_1 + n_2$ this will occur when $(n_1 - n_2)^2$ is least, that is, when $n_1 = n_2$, and there is a 1:1 sex ratio. This effect will be most important in small populations, for example, if a species is divided into small, relatively isolated groups.

In all situations considered above we see that the observed sex-ratios show a general agreement with those expected on the theory that the function of sex is to reduce inbreeding so far as possible. In the case of species which rear their young, a further reason for expecting equality of sex ratio at the time of sexual maturity has been given by Fisher in terms of the effort required to raise offspring of the two sexes.

Evolution of the Sexual Differentiation

We now turn from the consideration of the final population sex ratio to the mechanisms through which sexual diversity and the numerical equality of the sexes may have evolved. A survey of the animal and plant kingdoms shows that sex-differentiation has vanished and probably arisen many times during evolution, and it is safe to assume that no two instances were quite the same.

Let us again assume that a species requires the fusion of two genetically diverse types of gametes for its continued existence, and that fertilization will be most successful if the reserve material is stored mainly in one type. Some barrier is required to prevent gametes from the same individual from combining, and this must be powerful, or it will be frustrated by the much greater probability of encounter of such gametes as compared with those from different individuals. Three possible devices are as follows: (1) Self-sterility mechanisms, for example, self-incompatibility. These are sometimes rather precarious arrangements, found in fungi, other plants, and a few animals. (2) Proterandry or protogyny, where individuals produce male and female sex products at different times in succession. (3) The differentiation of two types (males and females), one producing small gametes, the other large ones.

In certain fish and mosquitoes single allelic differences or possibly short regions in a chromosome seem to determine sex. In most other organisms which have been investigated there have evolved different and more complex schemes of genetical sex determination. One difficulty which such mechanisms encounter during their evolution is that of apportioning three genotypes *AA*, *Aa*, and *aa* to two sexes with equal frequency.

One theoretical possibility would be to have both homozygotes *AA* and *aa* of one sex, and the heterozygote *Aa* of the other sex. This has not been found, although Whiting and Whiting discovered a somewhat similar mechanism of sex determination in the hymenopteran *Habrobracon judlandis*; there, heterozygosity at a number of loci on different chromosomes results in female development. Males are normally haploid, but can also be produced by homozygosity at some of these loci.

Another possibility is that of a homogametic sex, genetically *aa*, and a heterogametic sex, *Aa*; in this context *A*, *a* can denote either allelic genes or complete chromosomes (*Y* and *X*). This kind of sex determination is of frequent occurrence. It could arise through the gradual elimination of the type *AA*.

Let us suppose that in a diploid population homozygous for gene *a*, and producing only small gametes, there arises a dominant gene *A* which results in larger gametes with more reserve material being formed. These gametes which will carry *A* will thus provide the zygotes with more reserves, and thus be advantageous; hence the gene *A* may be expected to spread through the population. This spread will, however, be checked by the difficulty of union between these large, relatively static gametes, so that the advantage will only remain while there are a sufficient number of small gametes also present. Hence this situation may be expected to lead to an equilibrium in which both dominant and recessive types persist, and most fertile unions are between large and small gametes. From such a mating system in which there is already a tendency toward sexual differentiation, natural selection may be expected to develop a more complete sexuality in the course of time.

Another conceivable origin of sex-differentiation would be the appearance of a dominant mutation *A* with the property that types *AA* and *Aa* produce large gametes, and *aa* produces small gametes. Let us further suppose that there is dissortative mating, that is, a preference for the mating types dominant × recessive rather than dominant × dominant or recessive × recessive. Then the calculations of Li suggest that even with only a moderate degree of dissortation, the population will approach an equilibrium in which the proportions of dominants and recessives are equal, and only a small proportion of the dominants are homozygotes *AA*. Now the structure of such a population differs only little from one with completely dissortative mating, with only the two types *Aa*, *aa*, present in equal numbers and only the single type of mating *Aa* × *aa*. This in turn suggests that it may be comparatively easy for selection to change from moderate to complete dissortation; this involves little disturbance to the population as a whole, but provides it with the advantages of sexual differentiation.

It seems worth noting, as pointed out to us by Prof. L. S.

Penrose, that a dissortative mating can lead to an equilibrium also when the homozygote *AA* is lethal, whereas without dissortative mating this would lead to a gradual elimination of the gene *A*. If, therefore, a dissortative mating system arises during the development of sexual differentiation, it becomes possible for the *AA* type to tend to lethality without destroying the system, thus emphasizing the differentiation into the two types *Aa* and *aa*. This can lead ultimately to situations like the human one, in which the *YY* chromosomal zygote is presumably lethal.

Other mechanisms of ensuring numerical equality of the sexes are of course conceivable. It should, however, be pointed out that most of the tendencies which have been postulated in this article do not apply to the primary sex ratio, but to numerical sex equality at maturity. It has been frequently claimed but also denied that the primary sex ratio, that is the sex ratio at conception in man, is very high and it is a fact that the secondary sex ratio, namely, that at birth, shows regularly a significant excess of males. However, the survival of females is greater through most of post-embryonic life and women reach a higher age than men. This situation results in near equality of the sexes during the reproductive ages.

8: Comments on the Two Preceding Papers

A. W. F. EDWARDS

In a recent article, Kalmus and Smith have attempted to explain why a sex ratio of about one-half is selectively advantageous. More recently, Bodmer and Edwards have published a paper on the same subject. The two approaches differ in the emphasis placed on different forms of selection. Kalmus and Smith believe that there are two main selective forces: one which requires the chance of encounter of a male and a female at the reproductive age to be a maximum, and hence that the sex ratio at this age should be one-half (as suggested by Crew); and the other that the genetic variance of a population should be as large as possible for a given population size, which also requires a sex ratio of one-half at sexual maturity. On the other hand, Bodmer and Edwards maintain that the most important selective force arises through the advantage of reproducing as efficiently as possible—that is, of making the maximum genetic contribution to future genera-

From *Nature*, 188 (1960), 360–961.

tions for a given amount of effort devoted to the bearing of young. As Fisher pointed out, a sex ratio of about one-half is the most efficient in this sense.

The fundamental difference between these two approaches lies in the mode of selection. The types of selection that Kalmus and Smith believe important can only operate *between* populations, whereas the latter, or Fisherian, type can only operate *within* a population. Thus Kalmus and Smith maintain that populations with sex ratios about one-half are common because they have replaced populations with other sex ratios, while Fisher's argument supposes that even an isolated population will evolve toward a sex ratio of one-half. It is thus clear that only in exceptional circumstances will a population be at a selective advantage over a competing population on account of its sex ratio since both will already have evolved a sex ratio of one-half.

If interpopulation selection were of paramount importance in determining the sex ratio, one would expect to have found that isolated populations, which are not subjected to it, frequently had sex ratios different from one-half, which has not been observed. Furthermore, in polygamous species in which a single male, after defeating his competitors, fertilizes a large number of females, there is no longer any reason for it to be advantageous if the chance of encounter of a male and a female is at a maximum: prolific reproduction is ensured only if there is a plentiful supply of females and the sex ratio is very low. The fact that the genetic flexibility of the population is then somewhat restricted is unlikely to outweigh the advantages of such a sex ratio in interpopulation selection, where the ability to breed large numbers is of particular importance. Nevertheless, the sex ratio in such species seems to be about one-half.

Kalmus and Smith concede the relevance of Fisher's theory when they state that "in the case of species which rear their young, a further reason for expecting equality of sex ratio at the time of sexual maturity has been given by Fisher." But this is a misrepresentation of what Fisher actually wrote: his argument refers not only to species which *rear* their young but also to species in which reproduction involves some expenditure of effort by the parents—that is, to all sexually reproducing species. Furthermore, as Bodmer and Edwards have shown, it is not equal numbers of the sexes at maturity that is expected but equal numbers at the end of the period of parental dependence.

It thus seems that Kalmus and Smith have not fully appreciated the force of Fisher's argument. When the two points of view are compared in detail there is little doubt that it is Fisherian selection that is primarily responsible for the observed numerical equality of the sexes in many species.

V

Sex

The nearly universal existence of the sexual cycle of meiosis and fertilization—although it has not yet played a conspicuous role in the controversy—is perhaps the most crucial evidence on the importance of group selection. Why this is so is clear from the paper by Smith, who is usually rather firmly on the other side of the controversy. Sexual reproduction must stand as a powerful argument in favor of group selection, unless someone can come up with a plausible theory as to how it could be favored in individual selection. And if group selection can produce the machinery of sexual reproduction, it ought to be able to do many other things as well. This conclusion will be difficult for some people to accept (it certainly is for me), and I suspect that efforts to account for sex will figure more and more in discussions of group selection.

Yet even if it were conceded that sexuality is a mechanism of long-term group benefit that could be produced only by selection at the level of competing groups, there remains the problem of what the long-range benefits really are. That this is already a controversial problem in itself is apparent from the final selections.

9: *The Origin and Maintenance of Sex*

J. MAYNARD SMITH

At the cellular level, sex is the opposite of reproduction; in reproduction one cell divides into two, whereas it is the essence of the sexual process that two cells should fuse to form one. In this essay I shall ask what selective forces were responsible for the origin of the sexual process, and by what selective process is it maintained. It is easier to ask these questions than to answer them; the fact that we cannot answer them with confidence is a challenge to evolution theory.

I was led to think about these questions after being involved in a controversy with Professor Wynne-Edwards on a quite different problem. It is Wynne-Edwards' thesis that animal population numbers are regulated by behavioral mechanisms which have evolved because they prevent the population from outrunning its food supply. Such mechanisms may involve individuals in refraining from breeding; they can therefore hardly confer a selective advantage on the individual, although

Original for this volume.

they may confer an advantage on the group to which it belongs. Wynne-Edwards therefore believes that these mechanisms have evolved by a process of "group selection," whereby gene frequencies change because some groups of related individuals are more likely to survive than others.

There are formidable difficulties for a population geneticist in any such explanation. This can be seen most easily from the following argument. Suppose that an individually harmful mutation occurs. This mutation can be eliminated from the population by a single selective death—i.e., by the death of the first individual to carry the mutation. In contrast, suppose a mutation occurs that is beneficial to the individual but harmful to the group. The mutation will spread to all members of the group, and can be eliminated only by the elimination of the whole group. Thus the maintenance of a characteristic favorable only to the group requires N times as many selective deaths as the maintenance of an individually favorable characteristic, where N is the number of individuals in a reproductively isolated group. If groups are large, the selective cost of maintaining an "altruistic" character will be prohibitive. It is therefore reasonable to attempt to explain the behavioral mechanisms described by Wynne-Edwards by selection acting at the level of the individual, and I think this can often be done.

There is however one property, that of sexual reproduction, which is almost universal, and for which the generally accepted explanation involves, implicitly or explicitly, a process of group selection. In its least precise form, this explanation states that sexual reproduction confers on a species a greater capacity for rapid evolutionary change, and consequently that when the environment changes, those species which reproduce sexually are more likely to survive. I do not doubt that this explanation is in some sense correct, but it raises more problems than it solves. In particular, if the advantages conferred by sex are long-term ones, conferred on a group as a whole, how could the complex genetic basis for sexual reproduction arise in the first place? And if the disadvantages of sexual reproduction, at least in multicellular bisexual organisms, are as great as they appear to be at first sight, why is not sexual reproduction more frequently lost?

Sex as an Evolutionary Advantage

The first step is to state more precisely why sexual species can evolve more rapidly. Evolution consists of changes in gene frequency. A gene frequency will not change under selection more rapidly in a sexual than in an asexual species; indeed, if the sexual species is diploid, some changes in gene frequency will occur much more slowly. Hence if only a single gene frequency is changing, sex is no advantage.

The advantages of sexual reproduction arise only when two or more genetic changes are being favored simultaneously. This was recognized by Fisher, who concluded that "the only groups in which we would expect sexual reproduction never to have been developed would be those, if such exist, of so simple a character that their genetic constitution consisted of a single gene." However, it seems to me that Fisher does not specify precisely the circumstances in which sexual reproduction is an advantage. Thus he writes "if . . . the mutation rates . . . are high enough to maintain any considerable genetic diversity, it will only be the best adapted genotype which can become the ancestor of future generations, and the beneficial mutations which occur will have only the minutest chance of not appearing in types of organism so inferior to some of their competitors, that their offspring will certainly be supplanted by those of the latter." In other words, Fisher argues that in asexual species most beneficial mutations occur in individuals not destined to have descendants in the distant future, whereas in sexual species any beneficial mutation can be incorporated into the genotype of distant descendants, and that by virtue of this difference the rate of evolution in sexual species is more rapid.

I believe this argument to be fallacious, because, oddly enough, Fisher did not do the necessary sums. Thus suppose that a haploid population occupies an environment which changes, so that at two loci the initially common alleles, *a* and *b*, are at a selective disadvantage to the initially rare alleles *A* and *B*. Let P_{ab}, P_{Ab}, P_{aB} and P_{AB} be the frequencies of the four genotypes. Evolutionary progress is measured by the rate of increase of P_{AB}, which is initially very small. It is shown in Appendix 1 that if initially $P_{ab} \,.\, P_{AB} = P_{aB} \,.\, P_{Ab}$, then this "independence relation" will be maintained throughout the evolutionary change. Now the effect of sexual reproduction is to bring the genotype frequencies into agreement with the independence relation. If however the genotype frequencies already obey that relation, sexual reproduction will not accelerate evolution.

Now if in an asexual population the alleles *A* and *B* initially owe their presence to the recurrent mutation, $a \rightarrow A$ and $b \rightarrow B$, reaching an equilibrium between mutation and adverse selection, then it is easy to show that the "independence relation" is satisfied. Thus in this simple case, in which according to Fisher's argument sexual reproduction ought to be an advantage, sex in fact makes no difference. Crow and Kimura have worked out the consequences of Fisher's argument quantitatively, and conclude that sexual species can evolve at rates many orders of magnitude greater than asexual ones. But they assume that favorable mutations are unique events, each type occurring once and once only. If this implausible assumption is dropped, their argument falls to the ground.

But suppose that there exist initially two environments, in one of which gene *A* is an advantage, and in the other of which gene *B* is an advantage. Populations will then evolve, one with P_{Ab} large and the other with P_{aB} large. Suppose now that a third environment becomes available for colonization by both populations (this could be a new area, or a transformation of one or both the existing environments). Then in the colonizing population $P_{Ab} \,.\, P_{aB} \gg P_{ab} \,.\, P_{AB}$, and sexual reproduction would enable P_{AB} to increase more rapidly than could happen in an asexual population.

In other words, if two genetically different populations are adapted to different environments, sexual reproduction makes possible the rapid evolution of a new population, carrying genes from both parental populations, and adapted to a third environment. Notice that the initial genetic adaptations were the result of natural selection. Thus another way of looking at the matter is to say that sexual reproduction makes it possible to utilize genetic variance generated by past natural selection to adapt rapidly to new circumstances. If existing genetic variance has been generated by mutation, as suggested in Fisher's argument, then sexual reproduction confers no advantage.

The Origin of Sex

It follows that sexual reproduction does confer a long-term advantage in enabling genes initially present in different individuals to be brought together in a single individual, but only if the "species" (in this context, the group of individuals between which genetic recombination can take place) is divided into populations genetically adapted to different environments. Sex in this sense long preceded mitosis and meiosis; the processes of transduction and transformation achieve essentially the same end.

Sexual reproduction requires first that DNA from different ancestors be brought together in the same cell, and second that there be some mechanism of genetic recombination. The latter seems always to depend on a process of pairing between identical, or at least very similar, lengths of DNA, and on some process functionally equivalent to breakage and reunion of DNA molecules. It seems therefore that the enzymes required for genetic recombination could not have evolved because of the advantages conferred by sexual reproduction, because these advantages would not have existed until all the necessary enzymes had been perfected. In fact the same enzymes are probably used in repairing damaged DNA. As is so often the case in evolution, an organ—in this case a group of enzymes—which ultimately performs one function evolved in the first place because it performed another.

Thus genetic recombination may have been a by-product of selection for DNA-repairing enzymes. But in any case, before genetic recombination could occur, a means had to exist to bring DNA from different ancestors together in a single descendant. In higher organisms, the obvious selective advantage to such mechanisms arises from hybrid vigor; two homologous lengths of DNA may each be deficient, but in different cistrons, and may therefore complement one another. Unfortunately for this argument, bacteria appear not often to utilize this particular advantage of diploids or heterokaryons. An alternative selective advantage for DNA transfer in bacteria has been suggested by Hayes. Organs such as *F* pili used in transferring DNA might in the first instance have developed under the instructions of viral DNA, which would thereby ensure its own transfer to a new bacterium. Only later would such organs be used to transfer bacterial DNA.

A sexual process involving a haploid-diploid cycle and meiosis, as found in eukaryotes, depended on the prior evolution of mitosis, and hence of centromeres, spindles, and centrioles. A theory of the origin of mitosis has been suggested by Sagan; what is important in the present context is that the relevant selective advantages were to the individual (or if Sagan is right, to the symbiotic pair, of which one provided the basal body of a flagellum, which evolved into both centromere and centriole) and not to the group. Once the machinery of mitosis had evolved, in organisms already possessing the enzymes needed for genetic recombination, the evolution of meiosis is not too difficult to understand. What is not clear is whether meiosis arose in organisms in which the main phase of the life cycle was haploid or diploid. It is possible that meiosis arose in an organism which could exist as a haploid, a heterokaryon, or a diploid. In the absence of meiosis the transformations open to such an organism would be:

$$\text{haploid} \rightleftharpoons \text{heterokaryon} \rightarrow \text{diploid}.$$

Heterokaryosis would evolve because of the advantages of hybrid vigor. But in heterokaryons there is no way of regulating the proportions of the two kinds of nuclei, and in cells with small numbers of nuclei there would be a constant danger of losing one or the other type. Diploidy, whereby the two sets of chromosomes of different ancestry share the same spindle, might therefore originate as a more stable way of propagating a particularly favorable heterozygous genotype. But it is still necessary to invoke the long-term evolutionary advantages of genetic recombination to explain the origin of meiosis.

The Sex Ratio

In eukaryotes with meiosis it is usual for there to be two sexes, and for each individual to have one parent of each sex. (These rules are not universal; for example, ciliates break the first and hymenoptera the second; the first rule is also broken in hermaphroditic animals and monoecious plants, which are discussed in detail later.) If these rules are obeyed, it is easy to show that natural selection will produce a sex ratio of unity. Thus suppose for example that there are more females than males. Then a male will have on the average more offspring than a female. Therefore a gene tending to cause individuals of either sex to have more male offspring, or tending to convert females into males, or to favor the survival of males at the expense of females, will increase under natural selection until the sex ratio is unity. The same argument applies in reverse if there are more males than females.

In microorganisms the situation is more complex, and the population genetics of "sex ratio" is not understood. For example in *E. coli* there are $F-$, $F+$, and Hfr types, according to whether the F factor is absent, present in the cytoplasm, or incorporated into the chromosome. Transformations between the types can be represented as follows:

$$F- \rightleftharpoons F+ \rightleftharpoons Hfr.$$

$F-$ are readily transferred into $F+$ by the transfer of an F factor, but the reverse transformation is rare and difficult to demonstrate. It is therefore puzzling that most *E. coli* outside laboratories are $F-$. The explanation is presumably that some $F-$ bacteria are at a significant selective advantage as colonizers of new habitats (if they enter a habitat already occupied by $F+$ bacteria they will be transformed). This in turn can be explained, since genetic recombinants are normally $F-$bacteria which have received chromosomal material from *Hfr* bacteria. If this interpretation is correct, it is an interesting illustration of the advantages of sexual recombination.

The Maintenance of Sex

In unicellular organisms, the disadvantages of sex are not great. The usual pattern is for asexual multiplication to be interrupted by sexual fusion only when conditions are severe and when continued multiplication would in any case be impossible. In view of this, and also of the fact that microorganisms commonly adapt to changed circumstances by evolutionary change as well as by individual physiological adaptation, it is not difficult to see why sexual processes, once evolved, should have been maintained.

But in multicellular organisms with separate male and female individuals the disadvantages of sex are severe. Suppose that in such a species, with equal numbers of males and females, a mutation occurs causing females to produce only parthenogenetic females like themselves. The number of eggs laid by a female, k, will not normally depend on whether she is parthenogenetic or not, but only on how much food she can accumulate over and above that needed to maintain herself. Similarly, the probability S that an egg will survive to breed will not normally depend on whether it is parthenogenetic. With these assumptions the following changes occur in one generation:

		Adults		Eggs		Adults in next generation
parthenogenetic	♀♀	n	⟶	kn	⟶	Skn
sexual	♀♀	N	⟶ (crossed)	$\frac{1}{2}kN$		$\frac{1}{2}SkN$
	♂♂	N	⟶ (crossed)	$\frac{1}{2}kN$		$\frac{1}{2}SkN$

Hence in one generation the proportion of parthenogenetic females increases from $n/(2N + n)$ to $n/(N + n)$; when n is small, this is a doubling in each generation.

It follows that with these assumptions, the abandonment of sexual reproduction for parthenogenesis would have a large selective advantage in the short run.

It is well known that asexual varieties of plants arise quite commonly, and that their distribution, geographical and taxonomic, suggests that they are successful in the short term but in the long term doomed to extinction. Asexual varieties are much rarer among animals, although they do occur. It is not clear why this should be. Some possible reasons for the comparative rarity of asexual reproduction are:

i) Meiotic parthenogenesis, followed by fusion of egg and polar body, or of the first two cleavage nuclei, is equivalent to close inbreeding. In naturally outbreeding species the decline in vigor caused by inbreeding might counterbalance the advantage of not wasting material on males. This argument does not apply to ameiotic parthenogenesis.

ii) In many mammals and birds, and some other animals, both parents help raise the young. In such cases parthenogenesis would usually be a disadvantage.

At first sight it seems that hermaphroditism, or monoecy in plants, eliminates the selective advantage of parthenogenesis. In a hermaphrodite species, no material is wasted on males, and no more resources need to be expended on sperm than are needed to fertilize the eggs produced. This argument I believe to be erroneous, at least in the case of hermaphrodites with external fertilization, for the following reasons.

In any species the number of eggs laid (or seeds produced) will be limited by some resource R. In a hermaphrodite the same individual will also produce sperm. It is reasonable to assume that the production of sperm will make demands on the same resource R, which must therefore be shared between eggs and sperm. The argument in the preceding paragraph amounts to saying that the major part of R will be devoted to eggs, only enough being devoted to sperm to ensure that the eggs are fertilized. This conclusion is an example of the use of what J. B. S. Haldane once referred to as "Pangloss' theorem"—that all is for the best in the best of all possible worlds. Unhappily, Pangloss' theorem is false. In this case it assumes that natural selection necessarily produces a result favorable to the species, regardless of selection at the individual level. This is not so, because individual selection is usually more effective than selection favoring one group or species at the expense of another.

In fact, it is shown in Appendix 2 that in hermaphrodites with external fertilization, or monoecious plants with compulsory cross-fertilization, the resource *R* will normally be divided equally between eggs and sperm, or ovules and pollen. In such cases hermaphrodites would on the average have only half as many surviving offspring as parthenogenetic females. However, the argument in the Appendix does not apply to hermaphrodites with internal fertilization, or to self-fertilizing hermaphrodites, because in these cases individual selection will favor a limitation of the amount of sperm or pollen produced to that needed to ensure the fertilization of the available eggs. The conclusions to be drawn therefore vary according to whether a group has internal or external fertilization, as follows:

i. In groups with external fertilization, hermaphroditism would not increase the reproductive potential of a species. It is the common mechanism of reproduction in land plants, presumably because it has the advantage that an individual can fall back on self-fertilization in the absence of near neighbors. It does not protect a species against the evolution of parthenogenesis.

ii. In groups with internal fertilization, hermaphroditism does increase the reproductive potential of a species. It may be for this reason that it has become the typical method of reproduction in plathyhelminthes and in gastropods. In the former of these groups it has proved to be a pre-adaptation to parasitism. It does protect a species against the evolution of parthenogenesis.

The argument in this section seems to lead to the conclusion that, except in the special cases of animals in which both parents care for the young, and of hermaphrodites with internal fertilization, metazoan animals would be expected to give rise frequently to parthenogenetic varieties. Since in fact this conclusion is false, the argument must leave something out of account. Ultimately what is left out of account is the long-term evolutionary advantage of sex. But the rarity of parthenogenetic varieties of animals suggests that this long-term selection acts, not by eliminating parthenogenetic varieties when they arise, but by favoring genetic and developmental mechanisms which cannot readily mutate to give a parthenogenetic variety. It is not clear how this has been achieved.

APPENDIX 1

The Rate of Evolution in Sexual and Asexual Species

Consider first the evolution of an asexual haploid population varying at two loci, as follows:

genotype	ab	Ab	aB	AB
fitness	1	$1+K$	$1+k$	$(1+K)(1+k)$
frequency (generation n)	P_{ab}	P_{Ab}	P_{aB}	P_{AB}

Then if P'_{ab} etc., are the frequencies in generation $(n:1)$:

$$P'_{ab} = P_{ab}/T$$

$$P'_{Ab} = P_{Ab}(1+K)/T$$

$$P'_{aB} = P_{aB}(1+k)/T$$

$$P'_{AB} = P_{AB}(1+K)(1+k)/T$$

where

$$T = 1 + KP_{Ab} + kP_{aB} + (K + k + Kk)P_{AB}.$$

Hence, if

$$P_{ab} \cdot P_{AB} = P_{Ab} \cdot P_{aB}$$

then

$$P'_{ab} \cdot P'_{AB} = P'_{Ab} \cdot P'_{aB}$$

Thus, if the "independence relation" for genotype frequencies is satisfied initially, it will be satisfied in subsequent generations. The relevance of this fact is that all that is achieved by sexual reproduction is the production of a population obeying the independence relationship from one initially not satisfying it. It follows that sexual reproduction is an advantage only to populations which initially fail to satisfy the relationship.

I have discussed this problem in greater detail elsewhere.

APPENDIX 2

Resource Allocation in Hermaphrodites

Suppose that the number of eggs, n, and of sperm, N, produced by a hermaphroditic individual are limited by a common resource, R, of which an amount a is required to produce an egg and an amount b to produce a sperm:

$$\text{Then } an + bN = R, \qquad (1)$$

where R is a constant.

The argument is unaffected if some part of the quantities a and b are used in dispersing, protecting, or nourishing the gametes.

Let the typical members of a population at any time produce

n_0 eggs and N_0 sperm. On the average, each typical member will have one surviving offspring as a female, and one as a male.

Now consider a mutant individual producing n' eggs and N' sperm. If each egg and sperm has the same chance of giving rise to a surviving offspring as those produced by a typical individual, then a mutant individual will produce n'/n_0 offspring as a female, and N'/N_0 offspring as a male. (This conclusion will not hold for a self-fertilizing hermaphrodite, or for hermaphrodites with internal fertilization, for which the fewer sperm are produced, the greater chance each sperm has of fertilizing an egg.)

Hence the total number of offspring produced by the mutant is

$$T = n'/n_0 + N'/N_0, \tag{2}$$

and substituting from (1),

$$T = \frac{R - bN'}{R - bN_0} + \frac{N'}{N_0}.$$

$$\therefore dT/dN' = \frac{1}{N_0} - \frac{1}{R/b - N_0}. \tag{3}$$

Hence if $N_0 < \frac{1}{2}R/b$, dT/dN' is positive; i.e., mutations increasing N increase fitness, and selection will therefore increase N_0. Conversely, if $N_0 > \frac{1}{2}R/b$, selection will decrease N_0.

Thus there is a stable equilibrium when $N_0 = \frac{1}{2}R/b$, or when $bN_0 = an_0 = \frac{1}{2}R$. In other words, there is a stable equilibrium when the resource R is equally divided between eggs and sperm. The conclusion holds, however, only for hermaphrodites with external fertilization.

10: *Evolution in Sexual and Asexual Populations*

JAMES F. CROW
MOTOO KIMURA

It has often been said that sexual reproduction is advantageous because of the enormous number of genotypes that can be produced by a recombination of a relatively small number of genes. The number of potential combinations is indeed great, but the number produced in any single generation is limited by the population size, and gene combinations are broken up by recombination just as effectively as they are produced by it. Furthermore, for a given amount of variability, the efficiency of selection is greater in an asexual population than in one with free recombination since the rate is measured by the total genotypic variance rather than by just the additive component thereof.

On the other hand, unless new mutations occur, an asexual population has a selection limit determined by the best existing genotype, whereas directional selection in a sexual population can progress far beyond the initial extreme, as has been demon-

From *The American Naturalist*, 99 (1965), 439–450.

strated by selection experiments. The purpose of this article is to compare sexual and asexual systems as to the rate at which favorable gene combinations can be incorporated into the population, considering the effect of gene interaction, mutation rate, population size, and magnitude of gene effect. Most of the material is not new, but the various ideas have not been brought together in this context and we have introduced some refinements.

HISTORICAL

The question was first discussed from the viewpoint in which we are here interested by Fisher and Muller. We shall follow mainly the argument given by Muller.

In an asexual population, two beneficial mutants can be incorporated into the population only if the second occurs in a descendant of the individual in which the first occurred. On the other hand, in a sexual population the various mutants can get into the same individual by recombination. Only if the mutation rate were so low or the population so small that each mutant became established before another favorable mutant occurred would the two systems be equivalent.

The situation is illustrated in Figure 1, adapted from Muller's original drawings. The three mutants, A, B, and C are all beneficial. In the asexual population when all three arise at approximately the same time only one can persist. In Figure 1, A is better adapted than B or C (or perhaps luckier in happening to occur in an individual that for other reasons was more fit than those in which B and C arose) so that A is eventually incorporated. B is incorporated only after it occurs in an individual that already carries A, and C only in an individual that already carries both A and B. In the sexual system, on the other hand, all three mutants are incorporated approximately as fast as any one of them is in the asexual system.

The lower part of the figure shows a small population. Here the favorable mutants are so infrequent that one has time to be

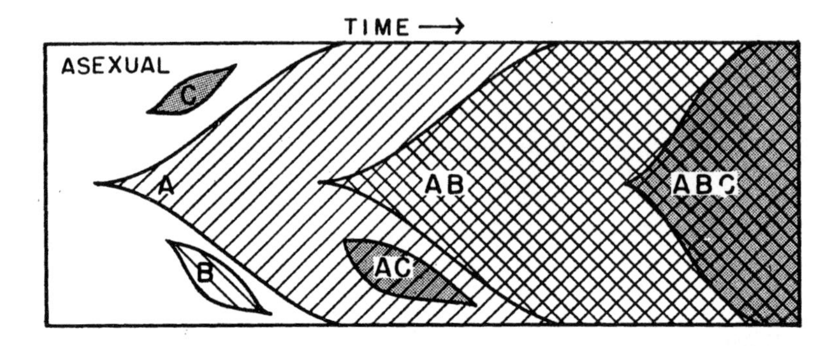

LARGE POPULATION

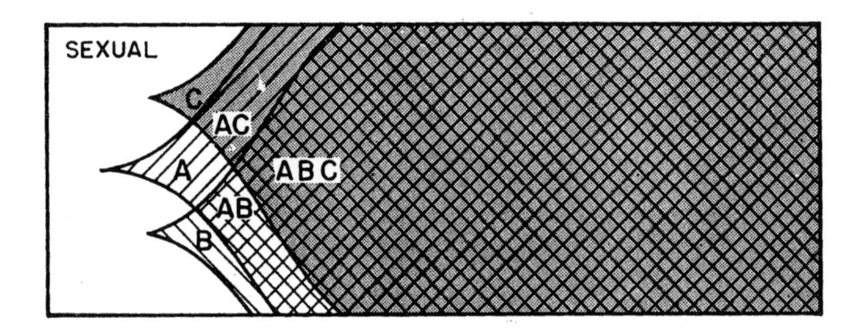

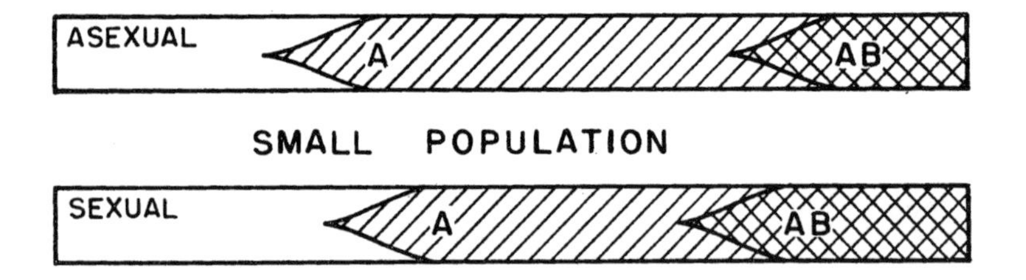

FIGURE 1: Evolution in sexual and asexual populations. The hatched and shaded areas show the increased number of mutant individuals following the occurrence of a favorable mutation. The abscissa is time. Modified from Muller (1932).

incorporated before another occurs. Thus there is no advantage to the sexual system.

In general, several favorable mutants arising at the same time can all be incorporated in a sexual system whereas only one can be without recombination. There is, of course, a high probability of random loss of even a favorable mutation in the first few generations after its occurrence. This problem has been solved by Fisher, but since the result is essentially the same in an asexual and sexual system it is irrelevant to the present discussion.

Muller's verbal argument was made quantitative in later papers. We have improved Muller's calculations slightly by taking into consideration the decelerating rate of increase in the frequency of a favorable mutant as it becomes common.

Mathematical Formulation

Consider first a population without recombination. We ignore the large majority of mutants that are unfavorable, for these are eliminated and thus are not incorporated into the population. Our interest is only in mutants that are favorable.

Let N = the population number
U = the total rate of occurrence per individual per generation of favorable mutations at all loci
g = the average number of generations between the occurrence of a favorable mutation and the occurrence of another favorable mutation in a descendant of the first
x = $1/U$ = the number of individuals such that on the average one favorable mutation will occur
s = the average selective advantage of a favorable mutant

Thus, g is the number of generations required for the cumulative number of descendants of a mutant to equal x, this being a number of such size that on the average one mutant will have occurred. Letting p_i be the proportion of individuals carrying the mutant gene in the ith generation, g is given by

$$x = Np_1 + Np_2 + \cdots + Np_g \qquad (1)$$

the summation being continued until there have been enough generations, g, to make the total number of mutant individuals equal to x. We assume that s is small, and therefore that p changes very slowly so that it is appropriate to replace addition by integration. This leads to

$$x = \int_0^g Np\,dt \qquad (2)$$

in the absence of recombination, p follows the logistic curve

$$p = \frac{p_0}{p_0 + (1 - p_0)e^{-st}} \qquad (3)$$

where p_0 is the initial proportion of mutants, as first shown by Haldane. If we start with a single mutant, $p_0 = 1/N$, and (3) becomes

$$p = \frac{1}{+ (N - 1)e^{-st}} \qquad (4)$$

Substituting this for p in (2) and integrating, we obtain

$$x = \frac{N}{s} \ln \frac{N - 1 + e^{sg}}{N}$$

or, rewriting and recalling that $x = 1/U$,

$$g = \frac{1}{s} \ln[N(e^{s/UN} - 1) + 1] \qquad (5)$$

In an asexual population one new mutant that will eventually be incorporated into the population arises every g generations. On the other hand, if reproduction is sexual, all the mutants that occur during this interval can eventually be incorporated. The number of mutants that arise per generation is NU, or in g

generations, NUg. Thus the ratio of incorporated mutations in a sexual population to that in an asexual population is $NUg:1$, or

$$\frac{NU}{s}\ln[N(e^{s/UN}-1)+1] \tag{6}$$

The favorable genes need not be mutants that are occurring for the first time. They may, for example, be genes brought in by immigrants or previously harmful mutants that have become beneficial because of a changed environment and already exist in the population at low frequencies.

Some numerical results are shown in Table 1. For example, if the selective advantage of a favorable mutation is 0.01 and the total rate of occurrence of such mutations is 10^{-8}, the

TABLE 1: *The Relative Rate of Incorporation of New Mutations Into the Population With and Without Recombination*

U/s	N						
	10^3	10^4	10^5	10^6	10^7	10^8	10^9
10^{-7}	1.0007	1.01	1.12	2.38	16.7	162	1.6×10^3
10^{-6}	1.007	1.09	2.15	14.4	139	1.4×10^3	1.4×10^4
10^{-5}	1.07	1.92	12.1	116	1.2×10^3	1.2×10^4	1.2×10^5
10^{-4}	1.69	9.75	92.6	922	9.2×10^3	9.2×10^4	9.2×10^5
10^{-3}	7.50	69.6	691	6.9×10^3	6.9×10^4	6.9×10^5	6.9×10^6
10^{-2}	46.7	462	4.6×10^3	4.6×10^4	4.6×10^5	4.6×10^6	4.6×10^7
10^{-1}	240	2.4×10^3	2.4×10^4	2.4×10^5	2.4×10^6	2.4×10^7	2.4×10^8

The ratio of the two rates is given in the body of the table. N is the population number, U is the total rate of occurrence of all favorable mutants, and s is the average selective advantage of such mutatants.

ratio U/s is 10^{-6}. As can be seen from the table, the advantage of recombination is negligible in a population of 10^3, but the rate ratio is 2.15 in a population of 10^5 and 1380 in a population of 10^8. If the selective advantage is smaller, the advantage of a sexual system is greater. Likewise, with a higher mutation rate the advantage is greater.

We have discussed mutants that were beneficial at the time of their first occurrence. Similar considerations are involved when a previously deleterious mutant type is rendered beneficial, such as by a change in the environment.

We do not intend to imply that the very high values in the lower right part of the table are realistic. Doubtless other factors become limiting. But the table does show the general trend and emphasizes the enormous advantage of an evolutionary system with recombination.

It is likely that the probability of a mutant being favorable is greater when the effect of the mutant is small. Thus, with small s, U tends to become larger. Since increasing U and decreasing s both have the effect of enhancing the advantage of recombination, the more that evolution proceeds by small micromutational steps, the greater the advantage of sexuality.

It is interesting that U and s enter the formula always in the form U/s, and never separately. This exact reciprocal dependence is understandable; for with slow selection the number of generations required for a given gene frequency change is inversely proportional to s. Thus a reduction in s means that proportionately more mutations will occur during the time that one is being incorporated.

Table 1 also shows that the advantage of recombination increases with an increase in the population size. In fact, with large populations the advantage is nearly proportional to the population number.

To summarize: The advantage of a reproductive system that permits free recombination is greatest for the incorporation of mutant genes with individually small effects, occurring at relatively high rates, and in a large population.

THE EFFECT OF GENE INTERACTION

So far we have been concerned only with mutant genes that are beneficial. We have also assumed that the combination of two mutant genes is more beneficial than either by itself; otherwise there would be no advantage of incorporating the second one.

The situation is quite different with some kinds of gene interaction. Where two or more mutants are individually harmful, but beneficial in combination, sexual reproduction may actually be disadvantageous.

The essential situation is clear with a haploid model, so we shall consider this simpler case. Suppose that the existing wild type in the population is genotype *ab*. The mutant types *Ab* and *aB* have fitnesses that, relative to *ab*, are reduced by the proportions s_1 and s_2. On the other hand, we assume that the double mutant has an enhanced fitness, greater than *ab* by a proportion t. The quantities s_1, s_2, and t are all taken to be positive.

Both single mutant types, *Ab* and *aB*, will be found in low frequency in the population, their exact numbers being determined by the ratio of their rate of occurrence by mutation to their rate of elimination by selection. The double mutant, *AB*, will occasionally arise, but infrequently.

However, once such a double mutant does arise, its fate will be quite different in a population with and without recombination. Ignoring the question of chance elimination during the early generations (which, as we said earlier, is not significantly different in the two kinds of populations), an *AB* double mutant in an asexual population will increase and eventually be incorporated at a rate determined by the value of t.

On the other hand, in a sexual population, an *AB* individual will ordinarily mate with an *ab* genotype, in which case the progeny will consist of all four genotypes in proportions depending on the amount of recombination. Only if the fitness of the *AB* type is great enough to compensate for the loss of *AB* types through recombination will this genotype increase. The relationships can be set forth as follows:

Genotype	ab	Ab	aB	AB
Relative fitness, w	1	$1 - s_1$	$1 - s_2$	$1 + t$
Frequency	$(1-x)(1-y)$	$x(1-y)$	$(1-x)y$	xy

$$\bar{w} = 1 - s_1x(1-y) - s_2(1-x)y + txy$$

$$\frac{\partial w}{\partial x} = y(s_1 + s_2 + t) - s_1$$

$$\frac{\partial w}{\partial y} = x(s_1 + s_2 + t) - s_2$$

These relationships assume that the two loci change independently under the action of natural selection, which is not strictly true unless $(1 + t) = (1 - s_1)(1 - s_2)$; but the formulae are approximately correct for unlinked loci when s_1, s_2, and t are small.

Gene A will increase when $\partial \bar{w}/\partial x$ is positive and decrease when this is negative. Therefore there is an unstable equilibrium at $y = s_1/(s_1 + s_2 + t)$. Below this value of y, x will decrease; above this value, x will increase. The situation is symmetrical for x and y by interchanging s_1 and s_2. Thus there is no way for the frequency of the AB type to increase unless it somehow gets past the equilibrium point. This problem was discussed extensively by Haldane.

The formulae are identical in a diploid population with complete dominance, on replacement of x by p^2 and y by q^2, where p and q are the frequencies of the recessive alleles at the two loci.

The situation is the familiar bottleneck frequently discussed by Wright. In his metaphor, the population is at one adaptive peak composed mainly of *ab* genotypes and there is no way for it to go to the higher peak composed mainly of *AB* genotypes without passing through a valley where *Ab* and *aB* types predominate.

There are several ways in which a sexual population might conceivably solve this problem. Some populations have several generations of asexual reproduction intervening between sexual generations. Another possibility would be strong assortative mating among the *AB* types; but the *a priori* probability of the genes that gave the increased fitness also producing the right type of mating behavior seems small indeed. Another possibility is random drift across the adaptive valley because of variable conditions or small effective population number, but this, as Wright has emphasized, would lead to a considerable lowering of fitness. Furthermore, Kimura has shown that the probability of joint fixation of two genes such as are being discussed here is very small, even in small populations. For example, in a population of effective size $N = 1000$, the single mutants with 1 per cent selective disadvantage and the double mutant with 5 per cent advantage, the probability of joint fixation is about $2.5\, p_0 q_0 \times 10^{-6}$, where p_0 and q_0 are the initial frequencies of the single mutants. The corresponding probability for completely neutral genes is $p_0 q_0$. Note that $p_0 q_0$ is a very small quantity. For individually deleterious but collectively advantageous mutant genes to have a reasonably high probability of joint fixation, the population must be so small that the inbreeding effect causes a serious effect on the viability.

In general, sexual reproduction can be a distinct disadvantage if evolution progresses mainly by putting together groups of individually deleterious, but collectively beneficial mutations. It seems to us that if this type of gene action were the limiting factor in evolution at the time sexual reproduction first evolved, sexual recombination might never have been "invented."

The Effect of Linkage

Two closely linked genes in a sexual organism can be quite similar to genes in an asexual organism as far as their relations to each other are concerned, for they may stay together for a great length of time. If r is the recombination frequency between two linked genes, they will stay together $1/r$ generations on the average before being separated by crossing over. This can easily be seen by noting that the probability that they will remain together g generations and separate in the next is $(1 - r)^g r$. Then the average number of generations during which they remain together is

$$\bar{g} = r + 2(1-r)r + 3(1-r)^2 r + 4(1-r)^3 r + \cdots$$
$$= r(1 + 2x + 3x^2 + 4x^3 + \cdots$$

where $x = 1 - r$. But $1 + 2x + 3x^2 + \cdots$ is the derivative of $1 + x + x^2 + x^3 + \cdots = 1/(1-x)$. Therefore

$$\bar{g} = r \frac{d}{dx} \left(\frac{1}{1 - x}\right) = \frac{1}{r}$$

Thus, two genes linked together with a recombination value of 0.1 per cent would remain linked on the average for 1000 generations before separating.

Consider again the earlier model where the four haploid genotypes, *ab*, *Ab*, *aB*, and *AB*, have fitnesses in the ratio $1:1 - s_1:1 - s_2:1 + t$, and assume that the amount of recombination between the loci is r. If the rare *AB* individual mates with an *ab* type, which will usually be the case, the proportion of *AB* progeny will be reduced by a fraction r because of recombination. However, the *AB* type will increase from these matings if the extra fitness of the *AB* type is enough to more than compensate for this; that is, if $(1 + t)(1 - r) > 1$, or $t > r/(1 - r)$.

The conditions for increase of *AB* genotypes in general are a little less stringent because some *AB* matings are with *AB*, *aB*, or *Ab* types and in these there is no effect of crossing over. Furthermore, *AB* types are being added by recombination from *Ab* × *aB* matings. Finally, $\bar{w}$, the average fitness is not 1, but slightly less. However, these do not change the direction of the inequality, so we can still say that a sufficient condition for the double mutant type to increase is $t > r/(1 - r)$. This is also the condition for increase in diploids, where t is now the advantage of the double heterozygote over the prevailing type.

In Wright's metaphor, the effect of linkage is to raise the valley between the two adaptive peaks and with extremely close linkage to provide a direct bridge.

For closely linked genes where r is small, the *AB* type will increase and ultimately become fixed if $t > r$. Thus, the closer the linkage, the greater the tendency to build up coadapted complexes—provided, of course, that such closely linked, mutually beneficial mutants occur. The extreme example is the high degree of functional interdependence within a cistron.

Coadaptation

We have seen that asexual organisms are in a better position than sexual species to build up coadapted complexes, except under conditions of close linkage. In an asexual population the mutants accumulate in a certain sequence; first we have mutant *A*, then *AB*, then *ABC*, and so on. In this case the effect of *B* in the absence of *A* is irrelevant; it may be beneficial or harmful, or simply be a modifier that is neutral in the absence of *A*. Of all the mutants that arise in the species after *A* has been incorporated, the one that is most likely to persist is the one that in the presence of *A* gives the greatest fitness. Therefore, there will be a tendency for combinations to be mutually coadapted, and these genes may be less beneficial or even harmful in other combinations. That is to say, they may well be what Mayr has called "narrow specialists."

In a sexual population, on the other hand, genes *A* and *B* are likely to be incorporated only if they are beneficial both individually and in combination. The type of gene that is most efficiently selected in a sexual population is one that is beneficial in combination with a large number of genes. We can only guess about the *a priori* distribution of gene interactions; but it is clear that in a population with free recombination the "good mixers" (that is, those having a large additive component) will be most efficiently selected.

The best opportunity to test these possibilities would be populations exposed to an entirely new environment. Drug resistance in bacteria and insecticide resistance in insects offer such a possibility. Chloramphenicol resistance in *Escherichia coli* is polygenic and has been analyzed by Cavalli and Maccacaro. During the selection for resistance the reproduction was asexual, though recombination was used later for analyzing the genetic basis of the resistance. Recombinants between resistant and susceptible strains were skewed in the direction of greater susceptibility, as were crosses between different resistant strains. The results, therefore, suggest considerable coadaptation with complementary action of the genes accumulated during the selection process.

DDT resistance in Drosophila is also polygenic and has been analyzed genetically. Analysis of variance of the contribution of various chromosomes to the resistance showed an almost complete additivity, as would be expected in a sexual species according to the view we have been discussing. Thus, at least in these two cases, there is good agreement with what our theoretical speculations would predict.

We should emphasize that the genetic variability in sexual populations that have had a long history of selection for the traits under consideration may not have a large additive component. The genes that act additively may already have been incorporated into the population so that those that remain in unfixed condition are the ones that are not responsive to selection; that is, they are genes with complex interactions. Thus it is not surprising if, in a stable natural environment or in an artificial population where selection has been practiced for a long time, the nonadditive components of variance predominate.

Haploidy versus Diploidy

The evolutionary advantages of recombination can be obtained in haploid as well as diploid species. Yet diploidy is the rule in a great many complex organisms and there must have been a regular trend of evolution from haploidy to diploidy.

At first glance it would appear that there is an obvious advantage of diploidy in that dominant alleles from one haploid set can prevent the deleterious effects of harmful recessive alleles in the other. However, when equilibrium is reached the situation is roughly the same in a diploid as in a haploid. In a haploid species the mutation load will equal the total mutation rate when equilibrium is reached. With diploidy the load will be somewhere between this value and twice this value, depending on the level of dominance. With any substantial heterozygous effect of deleterious recessive genes, the mutation load is nearly twice the mutation rate. So the effect of diploidy is generally to double the mutation load by doubling the number of genes. From this standpoint diploidy certainly offers no advantages, only disadvantages.

However, when the population has reached equilibrium as a haploid, a change to diploidy offers an immediate advantage. To be sure, when the population reaches a new diploid equilibrium the advantage is lost; but by then there is no turning back, for a return to haploidy would greatly increase the load by uncovering deleterious recessives. Thus, it is easy to see how diploidy might evolve from haploidy, even if the population did not gain any permanent benefit therefrom.

On the other hand, there are some other possible advantages of diploidy, of which we shall mention two. One that has frequently been suggested is the possibility of overdominance. To the extent that the heterozygote is fitter than either homozygote at some loci there is an advantage of diploidy, provided the average fitness of the diploid population is enough greater than the haploid to compensate for the greater mutation load.

A second possible advantage of diploidy is the protection it affords against the effects of somatic mutation, a possibility that also occurred independently to Muller. The zygote in a diploid species or the post-meiotic cell from which the organism develops in a haploid species might have approximately the same fitness at equilibrium, but the effects of somatic mutation would be quite different. If the soma were large and complicated, as in higher plants and especially animals, a diploid soma might provide a significant protection against the effects of recessive mutations in critical organs.

The Evolution of Sexuality

The development of sexual reproduction confers no immediate advantage on the individual in which this occurs. In fact, the result is far more likely to be deleterious. The benefit is only to the descendants, perhaps quite remote, and to the population as a whole. Thus, it seems likely that the selective mechanism by which recombination was established was intergroup selection. Fisher goes so far as to suggest that sexuality may be the only character that evolved for species rather than for individual advantage.

On the other hand, despite the great evolutionary advantages of sexual reproduction, there are immediate advantages in a return to asexual reproduction. An advantageous type whose recombinant progeny were disadvantageous would have an advantage for its immediate descendants by developing an asexual mode of reproduction, other things being equal. In diploids there is the additional advantage of fixing heterotic combinations.

This all accords with the conventional belief that sexuality developed very early in the evolution of living forms and is therefore found in all major groups; but that numerous independent retrogressions to vegetative reproduction continue to occur, conferring an immediate advantage but a long time evolutionary disadvantage.

Summary

In an asexual population two favorable mutants can be incorporated into the population only if one occurs in a descendant of the individual in which the other occurred. In a sexual population both mutants can be incorporated through recombination. A mathematical formulation is given of the relative rates of incorporation of the new mutations with and without recombination. Recombination is of the greatest advantage when the double mutant is more advantageous than either single mutant, when the mutant effects are small, when mutations occur with high frequency, and when the population is large.

On the other hand, for the incorporation of individually deleterious but collectively beneficial mutations, recombination can be disadvantageous. Close linkage has effects similar to those of asexual reproduction. Experimental data on DDT resistance in Drosophila and chloramphenicol resistance in bacteria are cited showing greater development of coadaptation in an asexual system.

The evolution of diploidy from haploidy confers an immediate reduction in the mutation load by concealment of deleterious recessives, but this advantage is lost once a new equilibrium is reached. Thus the development of diploidy may be because of an immediate advantage rather than because of any permanent benefit. On the other hand, there are other possible advantages of diploidy, such as heterosis and protection from somatic mutations.

11 : *Evolution in Sexual and Asexual Populations*

J. MAYNARD SMITH

It was argued by Fisher that sexual reproduction is the only characteristic of living organisms which owes its presence to the fact that it favors the survival of groups rather than of individuals. It is therefore important to understand precisely what are the advantages conferred on a group by sexual reproduction.

The orthodox answer to this question is that it accelerates adaptation to a changing environment because it makes it possible to combine in a single descendant mutations originally occurring in distinct individuals. This argument has recently been made quantitative by Crow and Kimura. They conclude that sexual reproduction can accelerate evolution by many orders of magnitude and that its effects are greatest when the evolving population is large, the frequency of beneficial mutations high, and the selective advantages small.

I believe these conclusions to be wrong. My reasons can most

From *The American Naturalist*, 102 (1968), 469–473.

easily be illustrated by a counterexample, in which according to the orthodox view sexual reproduction ought to be an advantage, but in which in fact it makes no difference. A comparison of this example with the treatment of Crow and Kimura shows that different assumptions have been made. The validity of the alternative assumptions, and the circumstances in which sexual reproduction is an advantage, are then discussed.

A Counterexample

The simplest situation in which sexual reproduction should, according to the orthodox view, acelerate evolution is that in which selection favors two alleles at different loci, both initially rare. Consider therefore the evolution of a haploid population varying at two loci. Initially the population inhabits an environment in which alleles a and b confer the greatest fitness, and are therefore the common alleles at their respective loci. The environment then changes, so that alleles A and B confer the greatest fitness. How far will sexual reproduction accelerate evolution from genotype ab to genotype AB?

We need first to find the equilibrium frequencies of the four genotypes before the change of environment. Let these frequencies, and the fitnesses of the four genotypes, be

Genotype	ab	Ab	aB	AB
Fitness	1	$1 - H$	$1 - h$	$(1 - H)(1 - h)$
Equilibrium frequency	P_{ab}	P_{Ab}	P_{aB}	P_{AB}

Let the mutation rate from a to A be μ_A, and from b to B be μ_B. Then after selection and mutation have acted, the relative proportions of the four genotypes are:

$$ab: \quad P_{ab}(1-\mu_A)(1-\mu_B);$$

$$Ab: \quad [P_{ab}\mu_A(1-\mu_B) + P_{Ab}(1-\mu_B)](1-H);$$

$$aB: \quad [P_{ab}\mu_B(1-\mu_A) + P_{aB}(1-\mu_A)](1-h);$$

$$AB: \quad [P_{ab}\mu_A\mu_B + P_{Ab}\mu_B + P_{aB}\mu_A + P_{AB}](1-H)(1-h).$$

Since at equilibrium these proportions do not change, P_{Ab}/P_{ab} is constant from generation to generation, and hence:

$$P_{Ab}(1-\mu_A)(1-\mu_B) = [P_{ab}\mu_A(1-\mu_B) + P_{Ab}(1-\mu_B)](1-H).$$

If we assume that H and h are small, but are large compared to the mutation rates, this reduces to

$$\left.\begin{aligned} P_{Ab} &= P_{ab}\mu_A/H \\ \text{and similarly} \quad P_{aB} &= P_{ab}\mu_B/h \\ \text{and} \quad P_{AB} &= P_{ab}\mu_A\mu_B/Hh. \end{aligned}\right\} \qquad (1)$$

From (1) it follows that

$$P_{ab}P_{AB} = P_{Ab}P_{aB}. \qquad (2)$$

If (2) is to be satisfied, and if the frequencies of genes A and B are p_A and p_B, respectively, it can be shown that the frequencies of the genotypes *ab*, *Ab*, *aB* and *AB* are $(1-p_A)(1-p_B)$; $p_A(1-p_B)$; $(1-p_A)p_B$; and $p_A\,p_B$, respectively. These frequencies arise as a result of recurrent mutation and selection but are the same as would be produced by a single generation of sexual reproduction.

We will now consider the changes in an asexual species after the environment has changed. Let the fitnesses be:

Genotype	*ab*	*Ab*	*aB*	*AB*
Fitness	1	$(1+K)$	$(1+k)$	$(1+K)(1+k)$

Let p_{ab}, p'_{ab} be the frequencies of *ab* in two successive generations, and similarly for the other genotypes. If the mutation rates are small compared to K and k, then:

$$\left.\begin{aligned} p'_{ab} &= p_{ab}/T \\ p'_{Ab} &= p_{Ab}(1 + K)/T \\ p'_{aB} &= p_{aB}(1 + k)/T \\ p'_{AB} &= p_{AB}(1 + K)(1 + k)/T. \end{aligned}\right\} \qquad (3)$$

Where $T = 1 + Kp_{Ab} + kp_{aB} + (K + k + Kk)p_{AB}$. From (3), $p'_{Ab}p'_{aB} = p_{Ab}p_{aB}(1 + K)(1 + k)/T$, and $p'_{ab}p'_{AB} = p_{ab}p_{AB}(1 + K)/(1 + k)/T$. Hence, if $p_{Ab}p_{aB} = p_{ab}p_{AB}$, then

$$p'_{Ab}p'_{aB} = p'_{ab}p'_{AB}. \qquad (4)$$

But it has already been shown [equation (2)] that this independence relationship is satisfied initially, and hence from (4) it will be satisfied throughout the evolution from *ab* to *AB*.

It is now apparent that sexual reproduction would not alter the rate of evolution, since all that sexual reproduction can do is to restore this independence relationship.

The Reason for the Discrepancy

Why is it that Crow and Kimura conclude that sexual reproduction can accelerate evolution, whereas in this counterexample discussed it makes no difference? The essential difference between their assumption and mine is that they regard mutation as a unique event, whereas I have treated it as a recurrent event. Before discussing the merits of these alternative assumptions it will be helpful to see how the uniqueness assumption affects their conclusions.

In their notation

N = the population number.

U = the total rate of occurrence per individual per generation of favorable mutations at all loci.

g = the average number of generations between the occurrence of a favorable mutation and the occurrence of another favorable mutation in a descendant of the first.

$x \quad = 1/U =$ the number of individuals such that on the average one favorable mutation will occur.

According to their argument, in an asexual population one new favorable mutation will be incorporated every g generations. In a sexual population, NUg favorable mutations will occur in that time, and all will be incorporated. Hence the relative rates of evolution will be as NUg:1.

The statement that NUg favorable mutations will be incorporated in g generations of a sexual species assumes that all these mutations are different. Clearly if particular mutations recur, the conclusion could be wrong by many orders of magnitude. More precisely, it is out by a factor equal to the number of times a particular mutation recurs in g generation in a population of N individuals.

The uniqueness assumption also leads to an underestimate of the rate of evolution of an asexual species. Thus it is assumed that the first individual in a population to undergo a favorable mutation is the *only* individual to leave any descendants many generations later, as is the first of his descendants to undergo a second favorable mutation, and so on. But if, for example, favorable mutations $a \rightarrow A$, $b \rightarrow B$, and $c \rightarrow C$ are all recurrent events, then any *abc* individual undergoing any of these mutations has a good chance of leaving *ABC* descendants many generations later.

Notice that the discrepancy introduced by assuming mutations to be unique is greatest when NU_g is greatest—that is, when the population is large, the rate of favorable mutations high, and the selective advantages small (and hence g large). These are precisely the conditions listed by Crow and Kimura as those in which sexual reproduction is most advantageous. Thus their conclusions depend critically on their assumption that mutations are unique events.

It is difficult to see what justification there could be for making this assumption. Presumably most favorable mutations are base substitutions. If so, a particular mutation would have a frequency of 10^{-8} to 10^{-9}, in contrast to the frequency of 10^{-5} to 10^{-6} for the frequency of all harmful mutations in a particular cistron. A frequency of 10^{-8} does not make an event unique, particularly when discussing the evolution of sex, which originated among micro-organisms: there can be 10^9 bacteria in a test tube.

The only mutations which can reasonably be thought of as unique events are those requiring two almost simultaneous events—that is, structural rearrangements. The frequency U of favorable structural rearrangements is likely to be so low as to wipe out the predicted advantage of sexual reproduction.

The Advantages of Sex

I am satisfied that the treatment of the problem given by Crow and Kimura is misleading. It does not follow that the "counterexample" I have considered is an adequate model either. What is needed is a general treatment similar to that given by Crow and Kimura, but omitting the uniqueness assumption. Pending such a treatment, it is interesting to see what conclusions emerge if we regard the counterexample as essentially correct.

The counterexample considers evolution at two loci, from $ab \rightarrow AB$. It is shown that if initially $P_{ab}P_{AB} = P_{Ab}P_{aB}$, then sexual reproduction confers no advantage. It is also shown that if the genetic variance in the initial population was generated by recurrent mutation, then this equation will be satisfied.

Suppose, however, there were initially two environments, in one of which *Ab* was the fittest (and therefore commonest) genotype, and in the other *aB* was the fittest genotype. Suppose now that a third environment becomes available for colonization from the two existing ones and that in this new environment *AB* is the fittest genotype. Then initially $P_{Ab}P_{aB} \gg P_{ab}P_{AB}$, and the increase in frequency of *AB* will be very much greater if reproduction is sexual.

In this simple case, then, if the genetic variance of a population has been generated by mutation in a uniform environment, sexual reproduction does not accelerate evolution when the environment changes. But if the genetic variance has been generated by selection favoring different genotypes in different environments, then sexual reproduction will accelerate adaptation to a new environment.

In other words, sexual processes are an advantage because they make it possible to bring together in one individual, not merely mutations which have occurred in different ancestors (because the same result can be achieved equally well by recurrent mutation), but different regions of DNA which have been programmed by natural selection in different ancestral populations in different environments.

SUMMARY

In a simple model case, the rates of evolution in an asexual and a sexual haploid population are compared. It is shown that, if the genetic variance of the population is generated by mutation in a uniform environment, sexual reproduction confers no advantage. But if the genetic variance has arisen because selection has favored different genotypes in different environments, then sexual reproduction will accelerate adaptation to a new environment.

These conclusions differ sharply from those reached by Crow and Kimura. The difference arises because those authors treated mutations as unique events, whereas they are here treated as recurrent events.

12 : *Evolution in Sexual and Asexual Populations: A Reply*

JAMES F. CROW

MOTOO KIMURA

In a recent article, J. Maynard Smith has questioned the validity of the concept, originally due to R. A. Fisher and H. J. Muller, that sexual reproduction is advantageous because it permits favorable mutants that occur in different individuals or lines to be recombined into one. In particular he questions the conclusions that we reached from a quantitative examination of Muller's model: that recombination is most advantageous when (1) the population is large, (2) the frequency of beneficial mutants is high, and (3) the individual selective advantages of the mutants are small.

Maynard Smith bases his main argument on a "counter example." He considers a two-locus, two-allele haploid case where the fitnesses of the normal, the two single mutants, and the double mutant are in the ratio $1:1-s:1-t:(1-s)(1-t)$.

From *The American Naturalist*, 103 (January–February 1969), 89–90.

The four types are assumed to be in equilibrium under mutation and selection. If, now, the environment changes so that the mutants become favorable, while the four fitnesses remain in geometric ratio, the course of evolution is the same whether there is recombination or not, and the double mutant is eventually fixed. We agree with the example.

This result, we believe, does invalidate a very minor part of Muller's original argument, but not the main part. Muller said that a second disadvantage of asexual reproduction is that the various favored genotypes must compete with each other, each slowing the evolution of the others until the best type finally wins. With recombination, on the other hand, the different mutants can interpenetrate and cooperate. Maynard Smith's example illustrates that Muller's argument is wrong, since there is exactly the same rate of change with recombination as without it. Curiously, it was exactly this example that in 1964 convinced Muller that the competition between genotypes is the same in sexual and asexual systems. Muller once scribbled a note on a postcard that he was glad to have the "competition hoax" demolished and hoped to write an article to this effect. For this reason we did not discuss the point in our paper. Muller's death came before he had a chance to write the article.

The main point of Muller's papers, and the one on which we elaborated, is the value of recombination in putting together rare advantageous genes that occur in separate individuals. The rate of gene substitution in a species with recombination exceeds that in a species without recombination by a factor roughly equal to the number of loci at which such substitutions are going on simultaneously. Fisher's statement that the relative advantage of sexuality is equal to the number of genes should say, rather, the number of *simultaneously evolving* genes, as he once said orally. Maynard Smith argues that, in a large population with recurrent mutation, doubly mutant individuals already exist, so that recombination is not necessary. We did not assume that favorable mutants are unique (in fact, even favorable mutants are usually lost by random processes, so a recurrent mutant has a better chance), but we did assume that they are so rare that multiple mutants would not exist in the same individual, even in a large population.

In our view, beneficial mutants are very rare. Doubly mutant individuals would occur at rates roughly the square of the single-mutant rates. Furthermore, if evolution occurs at the rate of a gene substitution every few hundred generations or oftener, there must be many genes being substituted at the same time, since it may require thousands of generations to incorporate a mutant with a very slight advantage. The starting point in an asexual population would then not be a double mutant, but an n-tuple mutant, which would be vanishingly rare even in the largest populations. So, we are unchanged in our view that the Muller-Fisher argument and our quantitative extensions have merit.

This is not to say, however, that the *only* advantage of recombination is in hastening the incorporation of rare mutant combinations. There may be similar advantages of recombination among frequent alleles, although we have not been able to find a satisfactory general formulation of the problem. Maynard Smith has suggested that sexual reproduction is advantageous in a heterogeneous environment where each single-mutant type is adapted to a particular niche and where a third niche, to which the double mutant is pre-adapted, opens up. Recombination is indeed of some advantage in this situation, but this seems to us to be a rather special example, and we would prefer a general theory that does not depend on a particular population structure.

Whether evolution depends mainly on the incorporation of rare favorable mutants or, alternatively, on a shifting balance of gene combinations where individual alleles are at intermediate frequencies is a matter that is still debated. If the former is true, we have, we think, given an important argument; if the latter is more often correct, we have given an argument that is unimportant, although it still may have been important at the time when sexual reproduction was first "invented."

For Further Reading

Some of the articles listed below would have been entirely suitable for reproduction in this book, but for one reason or another they were omitted. All are important to the controversy on group selection. The list is mainly of more recent contributions, some of which provide good introductions to older work.

ALLEE, W. C. *Animal Aggregations. A Study in General Sociology.* Chicago, Ill.: University of Chicago Press, 1938.

An early review of a variety of effects seen in groups of organisms rather than individuals. Allee did not produce a theory of group selection, but it is implied in some of his discussions, and some of his observations are difficult to explain without group selection.

BRAESTRUP, F. W. [review of Wynne-Edwards' "Animal Dispersion in Relation to Social Behaviour"] *Oikos,* 14:113–120, 1963.

One of the unfavorable reviews of Wynne-Edwards' book, with a number of thoughtful criticisms.

CHRISTIAN, JOHN J. [review of Wynne-Edwards' "Animal Dispersion in Relation to Social Behaviour"] *Quarterly Review of Biology,* 39:83–84, 1964.

One of the favorable reviews of Wynne-Edwards' book.

DUNBAR, M. J. *Ecological Development in Polar Regions. A Study in Evolution.* Englewood Cliffs, N. J.: Prentice-Hall, 1968.

An important thesis of this book is that the habitable environments of polar regions are new (recently relieved of continental glaciers) and that the ecosystems there are in an early stage of collective evolution. Some of the discussion implies selection at the ecosystem level, although theoretical treatment of this concept is explicitly reserved for the future.

GADGIL, MADHAV, and WILLIAM H. BOSSERT. "Life Historical Consequences of Natural Selection," *American Naturalist*, 104:1–24, 1970.

An important contribution to our understanding of the relations among growth rates, age of first reproduction, mortality, and other life history features in evolution. The authors recognize selection at the individual level only.

HALDANE, J. B. S. *The Causes of Evolution*. London: Longmans, 1932.

This book, by one of the founders of population genetics and the modern theory of evolution, includes an appendix that gives the first rigorous treatment of the natural selection of altruism.

HAMILTON, W. D. "Extraordinary Sex Ratios," *Science*, 156:477–488, 1967.

The most detailed treatment to date of the natural selection of sex ratio. An excellent introduction to the subject and a necessary point of departure for future work.

LACK, DAVID. *Population Studies of Birds*. London: Oxford, 1966.

A detailed account of the control of population density by environmental factors in several species of birds. A criticism of Wynne-Edwards' theory is provided in an appendix.

LEWONTIN, RICHARD C. "Selection in and of Populations," In *Ideas of Modern Biology*, edited by J. A. Moore, Natural History Press. 1965.

A well-reasoned account of the role of extinction in the evolution of life on earth. Also a summary of a curious sort of group selection in mouse populations.

MAYNARD SMITH, J. "What Use is Sex?" *Journal of Theoretical Biology*, in press.

Here the author elaborates further on some of the ideas expressed in his paper in this volume, and criticizes the theory advanced by the present editor (in entry under Williams, below), that sex can be a favorable reproductive strategy for an individual.

MULCAHY, DAVID C. "Optimal Sex Ratio in *Silene alba*," *Heredity*, 22:411–423, 1967.

The author finds evidence for group selection in sex ratios of a common weed.

MULLER, H. J. "Some Genetic Aspects of Sex," *American Naturalist*, 66:118–138, 1932.

The classic discussion of the evolutionary advantages of sexual recombination, and one of the first recognitions of sexuality as an argument for group selection.

WILLIAMS, GEORGE C. *Adaptation and Natural Selection: A Critique of Some Current Evolutionary Thought*. Princeton, N.J.: Princeton University Press, 1966.

This book develops arguments against group selection in several applications. It also advances reasons for believing that sexual reproduction is advantageous in individual selection, but does not explain how the disadvantage of genetic loss in meiosis can be overcome.

WRIGHT, SEWALL. "Tempo and Mode in Evolution: A Critical Review," *Ecology*, 24:415–419, 1945.

A book review that incidentally develops a model of group selection that depends on genetic drift in some of the selected populations.

WYNNE-EDWARDS, V. C. *Animal Dispersion in Relation to Social Behaviour*. Edinburgh: Oliver & Boyd, 1962.
The full account of Wynne-Edwards' theory of group selection and of the adaptive regulation of numbers by means of social organization.

Index